Einstein's Legacy

ML

9/95

EINSTEIN'S LEGACY

The Unity of Space and Time

Julian Schwinger

**SCIENTIFIC
AMERICAN
LIBRARY**

An imprint of Scientific American Books, Inc.
New York

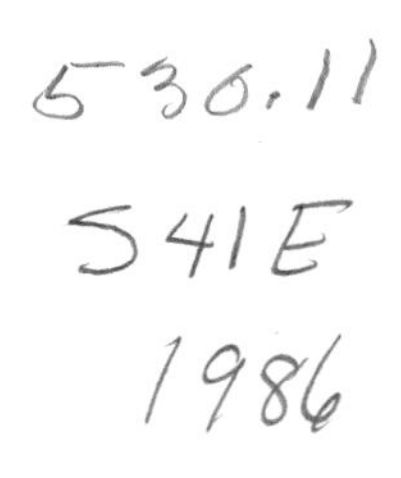

This is number 16 of a series.

Library of Congress Cataloguing in Publication Data

Schwinger, Julian, 1918–
Einstein's Legacy.

Bibliography: p.
Includes index.
1. Space and time. I. Title
QC173.59.S65S39 1985 530.1′1 85-19665
ISBN 0-7167-5011-2

Printed in the United States of America

Book design by Malcolm Grear Designers

Scientific American Library
An imprint of Scientific American Books, Inc.
New York

Distributed by W. H. Freeman and Company,
41 Madison Avenue, New York, New York 10010
and 20 Beaumont Street, Oxford OX1 2NQ, England.

1 2 3 4 5 6 7 8 9 0 KP 4 3 2 1 0 8 9 8 7 6

THIS BOOK IS DEDICATED TO

GEORGE ABELL,

WHO LOVED SCIENCE AND

HATED PSEUDOSCIENCE

CONTENTS

PREFACE

Somewhere I came across a statement, attributed to Einstein, emphasizing that the presentation of science to the general public must be as simple as possible, *but not more so.* That is the spirit in which I wrote this book on Relativity. Do not, however, assume that one day I sat down and began to write it. It is a longer story than that.

A number of years ago I was visited by a group of people then unknown to me: George Abell, Professor of Astronomy at UCLA, two members of the Open University of the United Kingdom, and two producers of the British Broadcasting Corporation. I learned that the Open University and the University of California had agreed to jointly sponsor several science series for presentation on television. The first such series was modestly titled "Understanding Space and Time," and my visitors had come to recruit me for the job of writing and presenting six programs on Relativity as part of that series. Their proposition fell on willing ears. I had earlier come up with a simple understanding of a few of the predictions of general relativity and therefore thought that I had something to contribute along such lines. (What I had more of was a lot to learn.) And, when George Abell confided in me that he was interested not only in the students of both universities but also in the vastly larger audiences of prime-time television, I was hooked.

In time, the programs were written (one BBC producer responded to my enthusiastic reading of a script with "I didn't understand a word"), were shot on location and in a London studio, and finally were packaged. George and I hosted an Extension course at UCLA, in which films of the programs were shown to an audience of some seven hundred faithful people. Meanwhile, discussions began with the local educational television station.

George proposed that we expand the scripts into a book to accompany the visual material that he was confident would shortly be available to large numbers of interested people.

We set to work, but at different paces. George was mainly occupied with the revision of one of his popular texts on astronomy. I faced a self-imposed deadline; I would soon be departing on a sabbatical leave. But, as I showed my successive chapters to George, he became increasingly unhappy. It seemed that we were addressing different audiences—he, the widest possible market; I, a smaller group of people, those who were willing to work a bit and thereby gain a much deeper insight. It was a relief to George when Scientific American Books proposed an independent printing of my six chapters; he assumed (correctly) that I would then be amenable to having him revise my contribution so that it fell into line with his intentions. Alas, it was not to be. George died without warning before anything materialized. Nor did the anticipated prime-time appearance of the television series ever occur. The local station had produced its own prime-time series, "Cosmos." To my knowledge, the only showing of "Understanding Space and Time" took place at the early hours reserved for events of no general public interest.

Plucking six chapters from the middle of a larger volume led inevitably to difficulties, both for my editors and for myself. So did the firm ideas I held about the tone of the book. Finally, however, an armistice was signed between the contending sides, and the result lies before you.

What was my general concept of the book? I had read enough popular expositions to gain the impression that, as the result of oversimplification, they consisted largely of disconnected statements, with little attempt to relate individual assertions to a more encompassing picture. That is inevitable when ordinary language is used rather than the concise symbolism that is the natural language of science. In speaking of reading the book of nature, Galileo said, "But it cannot be read until we have learned the language and become familiar with the characters in which it is written. It is written in mathematical language." Mathematics is often a cause for anxiety. But that need not be so when it appears in the context of science. Then, symbols—first introduced to present verbal statements in a convenient shorthand—become invested with their physical significance and acquire an intuitiveness and a potency far beyond the lifeless *x*s and *y*s of high school algebra. Think of the evocativeness of $E = mc^2$! As the latter reference emphasizes, Relativity—the special theory—comes already wrapped in symbols. And it is only through the use of symbols that the unity and the power of the concepts can be appreciated. I should be derelict if I gave the reader anything less.

The situation changes when we come to the general theory of relativity and its final synthesis in Einstein's gravitational equations. Here it seemed wiser to fall back on a verbal description if there were to be any success in providing the reader with some feeling for the reasoning that supports this singular achievement. These paragraphs have been signposted as "dangerous." The danger to which I allude is not that of derailing the reader—these lines can be skimmed or passed over without interrupting the narrative—but rather of so intriguing him by what is only suggested that he must, perforce, plunge deeper into the subject.

In counterpoint to the appearance of abstract symbols, these pages contain references to the concrete lives of the major participants in this drama. Here are no contributions to the history of science (my sources are all secondary, if, indeed, not tertiary) but rather reminders that science is a human activity, with practitioners who share the strengths and the weaknesses of all people, although not always in the same proportions.

It was surprising to find that such anecdotes could be controversial. For example, I had come upon the following Einstein story (H. Dukas and B. Hoffman, eds: *Albert Einstein: The Human Side.* Princeton University Press, 1979, p. 62):

In 1921, Einstein traveled to the United States with Chaim Weizmann—the future first president of Israel—who was a chemist. Concerning their Atlantic voyage, Weizmann said,

> Einstein explained his theory to me every day, and on my arrival I was fully convinced that he understood it.

I found this delectable; an editor sternly denounced it as "counterproductive." It does not appear in the text.

Publishing a book is a highly collaborative activity. My thanks go to all the members of the Scientific American Library team who pooled their talents to bring about this final product. And I cannot forget that *Einstein's Legacy* is the direct descendant of "Understanding Space and Time." Its BBC executive producer, Andrew Crilly, has shown me many kindnesses, including an (unsuccessful) attempt to convert my tennis swing into the wristiness needed for squash. Individual appreciations for supplying historical or scientific information are due Milton Anastos, Stephen Brush, and Robert Vessot. Words do not suffice to acknowledge the contributions of my wife, who not only supplied TLC and sympathy, but converted my impossible scribbles into typewriting and served as in-house editor.

Julian Schwinger
Los Angeles
September 1, 1985

Einstein's Legacy

1

A CONFLICT BROUGHT TO LIGHT

DRAMATIS PERSONAE

Issac Newton—whose name is celebrated in his three laws of motion and his theory of universal gravitation—once wrote that, if he had seen a little farther than others, it was because he had stood on the shoulders of giants. Albert Einstein also stood on the shoulders of giants—those of Isaac Newton and James Clerk Maxwell.

All three men, in their own lifetimes, were members of the Royal Society of London for Improving Natural Knowledge, or, simply, the Royal Society. This, the oldest scientific society in Great Britain, began informally in 1645 with weekly meetings of "divers worthy persons, inquisitive into natural philosophy and other parts of human learning, and particularly of what hath been called the New Philosophy or Experimental Philosophy."[1] Some of the first meetings were held at the Bull Head Tavern, Cheapside, one of several homes that preceded the present one at Carlton House Terrace, not far from Trafalgar Square. The Society was formally recognized in 1660 by Charles II, and it was incorporated by Royal Charter in 1662.

Early in 1672, Newton sent a description of his new reflecting telescope to the Royal Society; it was read to the members during the meeting at which he was elected a Fellow of the Society. In replying to the notice of election, Newton proposed to present "an account of a philosophical discovery, which induced me to the making of the said telescope . . . being in my judgement the oddest if not the most considerable detection which hath hitherto been made into the operations of nature."[2] Newton had physically separated white light into its component colors by passing the light

James Clerk Maxwell (1831–1879)

Sir Isaac Newton (1642–1727)

Sir Godfrey Kneller (1646–1723) painted this portrait in 1702.

through a glass prism.* This, and related discoveries, led him, the following year, to devise his theory of light being a stream of corpuscles, or particles.

Perhaps the most important contribution to science that the Royal Society has made in its three centuries of existence is its early role in publishing Newton's masterful account of his discoveries: *Mathematical Principles of Natural Philosophy—the Principia.* The book was licensed for publication in 1686 by the then president of the Royal Society, Samuel Pepys—amateur scientist, amateur musician, treasurer of the Royal Fishery, member of Parliament, Secretary of the Admiralty—best known today for his daringly honest, secret account of his times, the *Diary.*

Just as Isaac Newton dominated the scientific scene in the seventeenth century, so Albert Einstein dominates that of the twentieth century. World renowned for his introduction and development of the theory of relativity, Einstein became a foreign member of the Royal Society in 1921, the same year for which he received the Nobel Prize, although it was not awarded until 1922. The citation for that prize does not mention the theory of relativity explic-

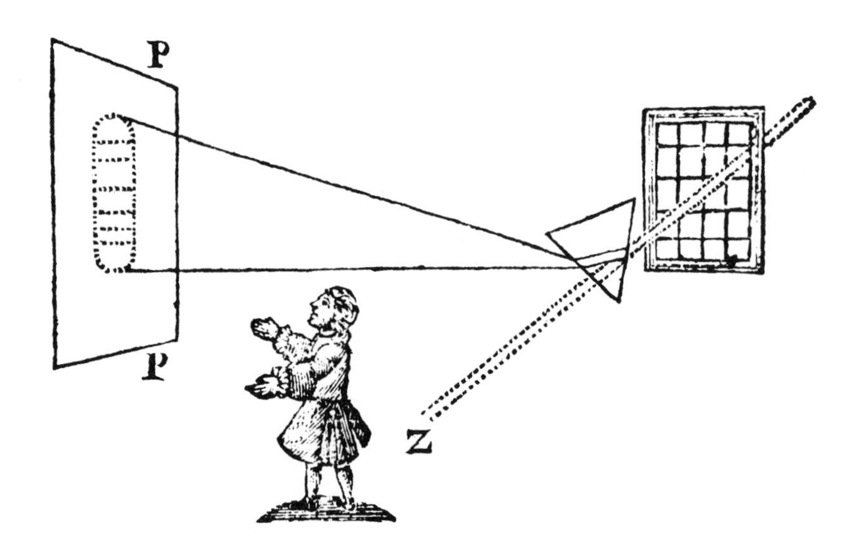

Newton: "Light is not similar or Homogenial, but consists of Difform Rays, some of which are more Refrangible than others"

This charming representation of Newton's discovery (taken from Voltaire's Eléments de la Philosophie de Newton *(1738)) shows a beam of sunlight that passes through a small hole in the window shutters, traverses a prism, and falls upon a screen where it appears spread out into the whole range of colors.*

*But Leonardo da Vinci (1452–1519) had been there before him: "If you place a glass of water on the windowsill so that the sun's rays will strike it from the other side, you will see colors formed in the impression made by the sun's rays that have penetrated through that glass and fallen on the pavement . . . ."[3]

itly; it reads "for his services to Theoretical Physics and especially for his discovery of the law of the photoelectric effect."

The latter refers to Einstein's role in reintroducing the particle nature of light, which Newton had proposed more than two centuries earlier. Against it was ranged the wave theory suggested by Christian Huygens (1629–1695) in 1690. Although there were doubters—Benjamin Franklin was one—Newton's authority had kept the particle theory in the forefront until the nineteenth century, when the wave theory won general acceptance (see Box 1·1). Einstein's discovery was not a return to Newton, however; the truth that ultimately emerged is more subtle than either of the two alternatives and transcends both of them. This part of Einstein's legacy—it concerns the laws of atomic physics—receives occasional mention in subsequent chapters. It is, however, outside the general framework of this book, which is focused on Einstein and relativity.

Einstein is a household word. Newton has his fan club.

Nature and Nature's laws lay hid in night:
God said, Let Newton be! and all was light.

ALEXANDER POPE *(1688–1744)*

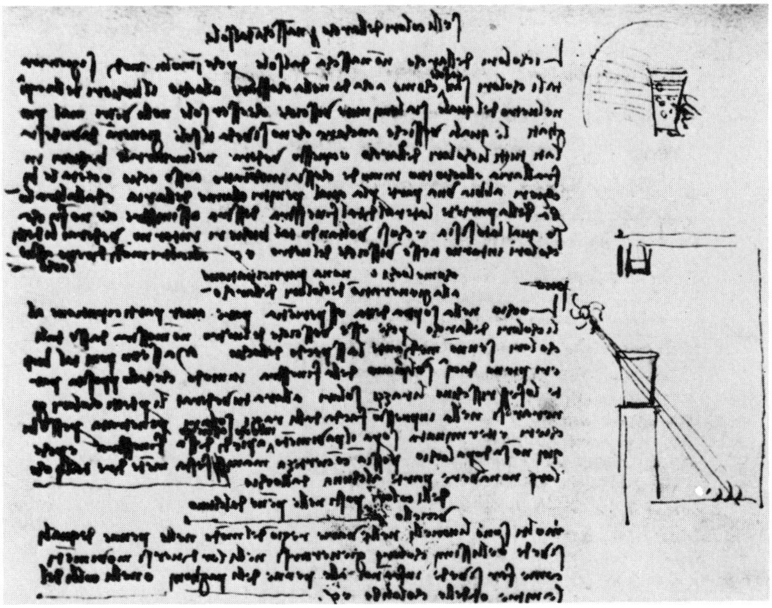

Leonardo da Vinci's drawing of his demonstration that sunlight can be decomposed into distinct colors

In this experiment, which preceded Newton's by more than a century, a glass of water produced the same effect as did Newton's prism. The accompanying description of Leonardo's experiment appears in his characteristic left-handed mirror writing.

BOX 1·1 ***Theories of Light***

Newton's corpuscular theory had the advantage, in the eighteenth century, of giving an obvious account of the straight-line propagation of light in a uniform medium. The theory could also account for the reflection of light from a surface and the refraction of light as it crosses the boundary between two optically different transparent media—air and water, for example. Newtonian mechanical concepts predicted that the speed of light would *increase* in an optically denser medium (here, water.) The wave theory of Huygens could also explain these phenomena but required that the speed of light *decreases* in an optically denser medium. But, before the crucial experiment measuring that speed was performed, another phenomenon tilted the balance between the confronting views.

The corpuscular theory states that any obstacle in the path of a light beam must produce a sharp shadow. Yet, even in Newton's time, it was known that shadows are not perfectly sharp, a phenomenon called *diffraction.* At the beginning of the nineteenth century, Thomas Young (1773–1829) introduced the wave concept of *interference* of light, to account for a striking example of diffraction, which is referred to as a *diffraction pattern.* You can see one for yourself.

Take two straight-edged cards and, with the edges in close proximity, hold them, very close to your eye, toward a strong light, blocking it out, except for what you can see through the slit between the edges. As the gap between the edges is narrowed, there will appear in the slit a succession of dark and bright lines parallel to the edges. Young's explanation for those lines was that waves passing through a narrow slit are *diffracted*—deviated from a straight-line path. (The phenomenon is commonplace with sound waves; out of sight is not out if earshot.) As a result, each point of the image is produced by waves coming from different parts of the slit. These waves travel different distances to an observer's eye, alternately canceling and reinforcing each other to produce dark and bright bands of light called *interference fringes.*

With the accumulation of such experimental evidence favoring the wave theory, it was an anticlimax when, in 1850, Jean Foucault (1819–1868) showed experimentally that light traveling in water *is* slowed down, not speeded up.

How, then, was it possible for Einstein to revive this discredited concept of light? By changing the laws of mechanics, particularly as they apply to light.

But James Clerk Maxwell (1831–1879), whose scientific accomplishments have had much greater effect on our daily lives, is comparatively unknown. Who was this man, and what did he do?

He was the last of a line of the well-to-do, land-owning Clerk family of Scotland; the Maxwell name had been added in order to retain lands acquired by marriage. The only surviving child of middle-aged parents, he was born in Edinburgh in the same summer in which Michael Faraday made an epochal discovery, one that Maxwell would later use as a cornerstone of his greatest achievement. The child soon displayed an omnivorous curiosity and a remarkable memory—he was *different*. Those characteristics, combined with a speech defect and shyness, led to a lonely existence at school, where he was the constant object of torment by his classmates. His stout resistance to this persecution was leavened by an irrepressible sense of humor. He survived and flowered. Later in life he remarked sadly, "They never understood me, but I understood them."

The early death of his mother, at age forty-eight, had put the boy's education into his father's hands. Although the doting father, John Clerk Maxwell, blundered badly in his initial choice of a tutor, who turned out to be brutal, James's later successes at school led John to take the boy to meetings of the Royal Society of Edinburgh. Results were not long in coming.

His first scientific paper, written when he was fourteen, was read for him before the Royal Society and published by that institution in 1846. In it he gives a method for constructing curves that are known as Cartesian ovals. It generalizes the way that an ellipse can be traced with a pencil by keeping taut a piece of string attached at two fixed points (the foci of the ellipse).

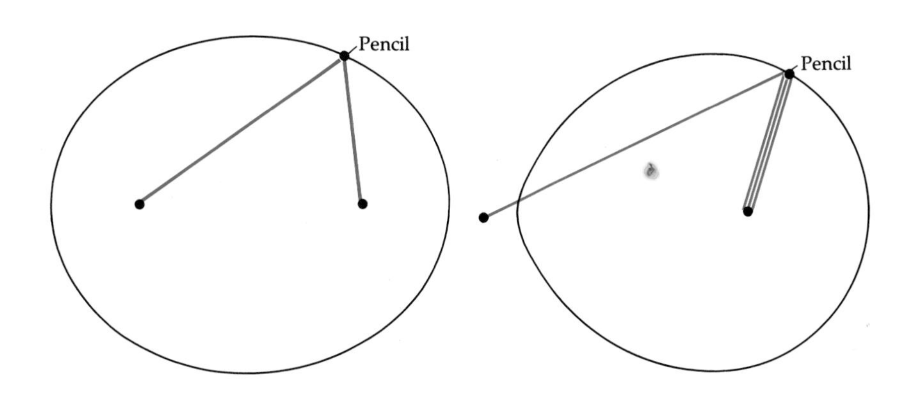

Drawing an ellipse and a Cartesian oval

Push two pins into a flat board covered by a sheet of paper. To these pins attach the ends of a string long enough to connect the pins loosely. To draw an ellipse (left), use a pencil to stretch the string taut and keep it that way as you trace a closed path around the pins; they mark the foci of the ellipse. For all points on the ellipse, the sum of the distances to the foci is a constant—the fixed length of the string. In a Cartesian oval (right), the string connecting a focus with the pencil will be folded back on itself a number of times, chosen independently for each focus. In this example, the somewhat longer string is attached to the right pin, runs around the pencil and around the pin again, and then loops around the pencil directly to the other focus. Here the fixed length of the string is the sum of the distance from any point on the oval to one focus (which is outside *the oval) and three times the distance to the other focus (within the oval).*

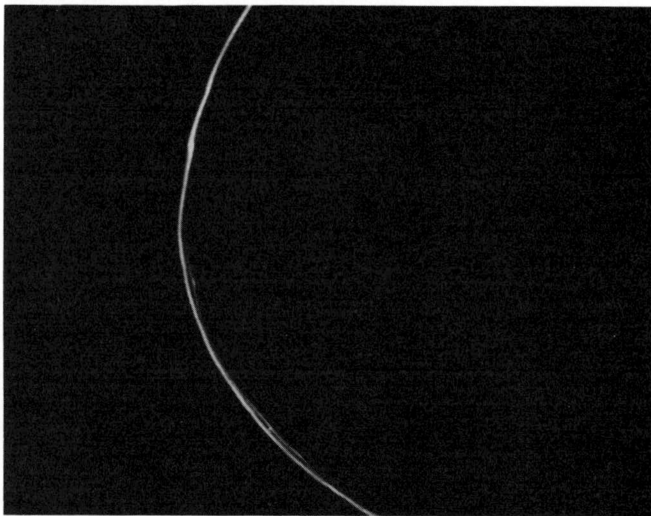

The main rings (left) and the braided ring (right) of Saturn.

The Voyager fly-by of Saturn in November 1980, guided perfectly by Newton's laws, left no doubt about the correctness of Maxwell's conclusion; the small bodies seem to be some form of ice. At the same time, Voyager disclosed totally unexpected structures in the rings. Maxwell would have been delighted. It is fitting that his name has been attached to some of these new features.

In 1855, when Maxwell was twenty-four, a competition was announced at Cambridge, where Maxwell was a Fellow of Trinity College. (Newton had also been there, almost two centuries before.) The prize was to be awarded for the best study of the rings of Saturn, focusing particularly on the question of their stability. Some seventy years earlier, the French astronomer Pierre Simon Laplace (1749–1827) had asserted that the rings are irregular solid bodies. After showing that a solid structure was either inherently impossible or contrary to observation, Maxwell demonstrated that the rings must be composed of very many small bodies. Happily, he won the prize.

Maxwell's work on Saturn's rings eventually directed his attention to another subject dealing with myriad small bodies: the molecular theory of gases. According to this theory, the pressure of a gas is produced by the collisions of the many tiny molecules against the walls that confine the gas. But molecules also collide with each other. One consequence of this is a resistance to flow, which is called *viscosity*. Maxwell proceeded to show that molecular viscosity would be independent of pressure. Then, as he was one of that rare breed of scientist—outstanding theorist *and* outstanding experimenter—Maxwell set out to prove this prediction by experiment. For two years, in the mid-1860s, Maxwell and his wife, Katherine Mary Dewar, made measurements of gaseous vis-

BOX 1·2 ***Maxwell's Demons***

All of us have heard of little people who could accomplish marvelous things: the leprechauns of Ireland, the Menahune of Hawaii. And there was the dwarf (or was it his mother?) who, in one night, built the towering Mayan pyramid at Uxmal, Yucatán. Maxwell uncovered a tribe of tiny beings who could do *impossible* things. They were "very small but lively beings incapable of doing work but able to open and shut valves which move without friction and inertia."[4] To give an example of their mischief, suppose that there are two similar volumes of gas at the same temperature, connected by such a valve, which, when open, allows the passage of the gas in either direction. Then one of the demons is likely to station himself gleefully by the valve and proceed to open and close it as follows. When he sees a particularly fast molecule approaching from one side, or a particularly slow one approaching from the other side, he manipulates the valve to let that molecule pass. When the situation is reversed, the valve is shut. In consequence of this transfer of energy from one side to the other, the temperatures of the two volumes become unequal, although no work has been performed. But, according to a principle known as the second law of thermodynamics,* that is impossible.

The way to exorcise the demon was found eventually. To operate the valve, the demon must *see* the approaching molecules. But at the beginning of his manipulations, when constant temperature—thermal equilibrium—prevails, radiation is moving uniformly in all directions, unaffected by the molecules. There is no contrast, no way to see the molecules. (A skier caught in a whiteout experiences this.) Well then, give the demon a flashlight. Ah, but that's a whole new ball game. Now work *is* being done in operating the flashlight. There is no objection to the demon producing a temperature difference—running a refrigerator—provided there is a compensating change in something else.

---

*"No process is possible in which the *sole result* is the transfer of energy from a cooler to a hotter body" is the statement of it given in *The Second Law,* by P.W. Atkins (Scientific American Books. 1984), p. 25.

cosities at different pressures. In the course of these experiments, carried out in their London home, they had to maintain various constant temperatures in their work room, sometimes with roaring fires, sometimes with a vast amount of ice—a far cry from today's multimillion dollar laboratories. The results vindicated Maxwell's molecular theory.

Maxwell's crowning achievement was his unification of electricity and magnetism in the electromagnetic theory of light. To understand this accomplishment, it will be helpful to review some of the experimental and theoretical developments after the time of Newton.

LIGHT

Newton had shown that the gravitational force of attraction between two massive bodies, at a distance apart that is large compared with their individual sizes, is directed along the straight line connecting the bodies with a strength that varies in inverse proportion to the square of that distance. This inverse-square law of force asserts, for example, that halving the distance quadruples the strength. The experimental investigations of Charles Augustin Coulomb (1736–1806) and of Henry Cavendish (1731–1810),* in the second half of the eighteenth century, revealed that the electrical force between charged bodies also has these characteristics, except

Inverse square law

The inverse square law *not only governs gravitational, electrical, and magnetic forces, but also describes the intensity of light emitted from a small light source, the flow of heat from a small heat source, and so forth. What all these examples have in common is that a constant amount of some quantity—call it total flux—is distributed uniformly over a sphere. The area of a sphere being proportional to the square of its radius, the amount of flux that is distributed over a unit area decreases as the total area increases, varying in inverse proportion to the square of the radius. This is illustrated here for a fixed fraction of a sphere that has been produced by drawing a cone from the center. The three surfaces are at distances in the proportions 1:2:3.*

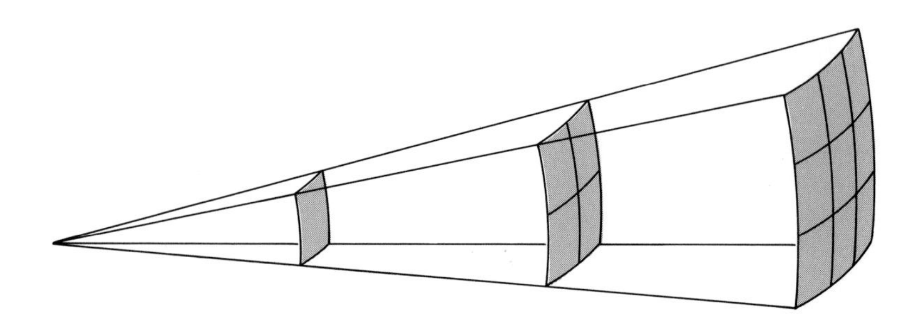

*Apart from attending the weekly meetings of the Royal Society, Cavendish was a determined recluse. His electrical experiments remained unpublished until, about a century later, Maxwell edited and published them in the last five years of his life. This established Cavendish's priority in the discovery of the electric-force law.

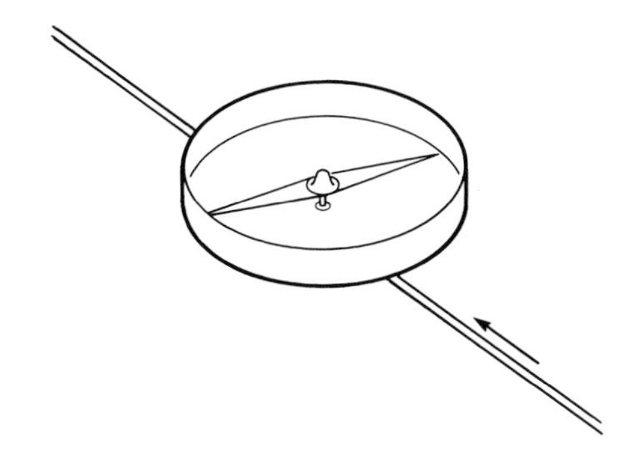

that the force can be either repulsive or attractive: like electric charges repel, unlike charges attract. Magnetism behaves similarly, with North and South poles playing the roles of positive and negative electric charge. Therefore, early in the nineteenth century, there was no reason to question the universality of the Newtonian pattern of forces.

Then, in 1820, the Danish physicist Hans Christian Oersted (1777–1851) broke the news that a *magnetic* compass needle, placed near a wire carrying a current (i.e., a flow of *electric* charge), is influenced by that proximity. Electricity and magnetism are related!* But the compass needle was neither attracted nor repelled; rather, the needle aligned itself perpendicularly to the current. Here was something new.

And yet, this new force could be squeezed into the Newtonian pattern. Almost immediately (1820), the French physicist André Marie Ampère (1775–1836) discovered that two wires carrying currents also exert forces on each other (which led him to the hypothesis that all magnetism is attributable to the flow of electric charge). As a simple example of such forces, consider the currents carried by two long parallel wires: if the currents flow in the same direction, there is an attractive force between the wires; opposite flows

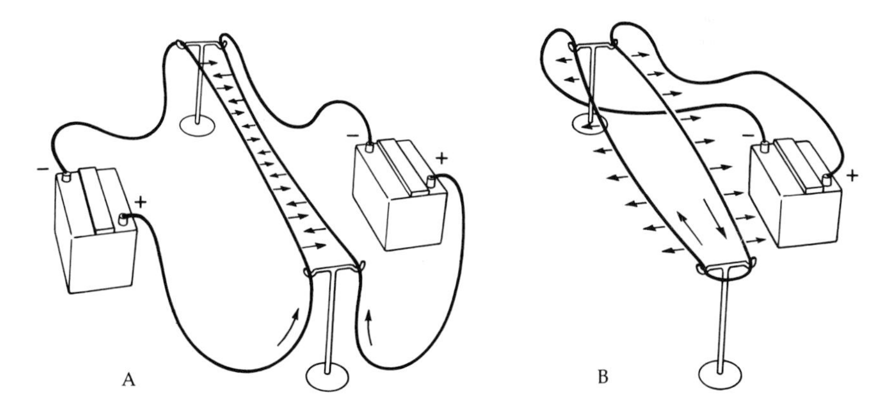

Parallel wires carrying currents

Parallel wires that carry currents in the same direction (A) are pulled together; those that carry currents in opposite directions (B) are pushed apart.

*This was not entirely unexpected. In 1681, a ship heading for Boston was struck by lightning. It was then discovered that the poles of the compass needle had become reversed. However, until Benjamin Franklin flew his kite (1752), it was not known that a lightning bolt is an electric current.

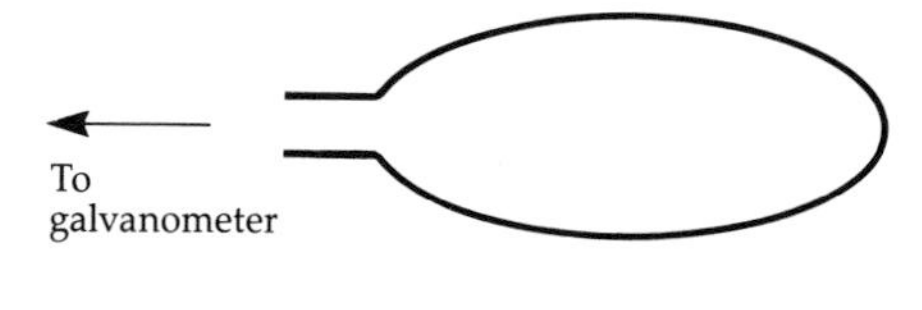

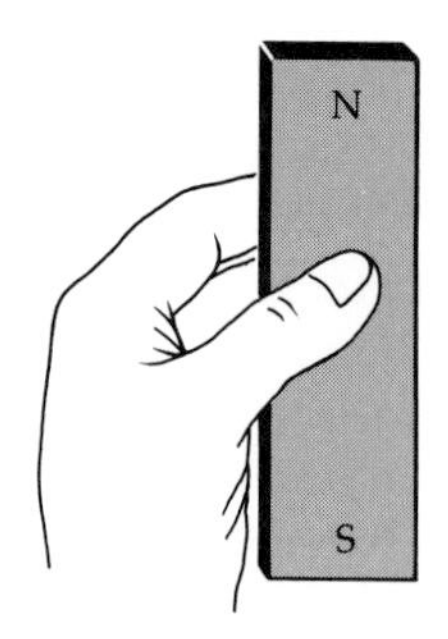

Magnetic induction

When a bar magnet is thrust through a wire loop (a coil of wire is more practical), an electric current is produced in the wire, ceasing as the motion stops. The rapid withdrawal of the magnet produces a similar flow of electric charge, but in the opposite sense. The electric current is detected by a galvanometer, which is basically a little magnet suspended near wire coils. The presence of an electric current in those coils is signaled by a deflection of that magnet.

Using the language of fields (see text), we can say that the motion of the bar magnet produces a changing magnetic field in the vicinity of the wire loop. The consequent flow of electric charge implies that an electric field has been generated. Faraday's discovery in a nutshell: A changing magnetic *field creates an* electric *field.*

produce a repulsion. This force varies inversely with the distance between the wires. That is not a contradiction: the lengths of the wires are *not* small compared with the distance between them. Indeed, Ampère was able to show that in general such forces can be considered to be built up from elements of force between small segments of wire. These elements of force are directed along the straight lines between the segments and vary in inverse proportion to the *square* of the distance between segments. Apart from the added complication that the elements of force also depend on various angles—that between the directions of the segments and those made with the lines connecting them—this is fully in the Newtonian spirit.

Enter Michael Faraday (see Box 1·3). After ten years of investigations, he discovered in 1831 that electric currents are induced in a conducting circuit by *changes* in a nearby magnet, whether produced by moving it or by altering its strength, which is easily done with an electromagnet. However, what is important here is not this discovery in itself—indeed, Ampère's force could be used to describe this phenomenon of current induction—but the way in which it turned Faraday's thoughts toward a new and ultimately revolutionary direction.

BOX 1·3 ***Michael Faraday***

Michael Faraday laid the foundations for the laws of electromagnetism at the Royal Institution in London, which had been founded in 1799 by Count Rumford (see Box 3·1) as "an establishment in London for diffusing the knowledge of useful mechanical improvements," and to "teach the application of science to the useful purposes of life."

In 1801, Rumford hired chemist Humphrey Davy (1778–1829), then aged twenty-two but already known for his discovery of the properties of nitrous oxide, or "laughing gas." It was Davy who realized Rumford's intent by combining original research with polished popular exposition and in the process made science fashionable. He began the 1805 lectures for the general audience with these words:

> The love of knowledge and of intellectual power is a faculty belonging to the human mind in every state of society; and it is one by which it is most justly characterized—one most worthy of being cultivated and extended.[5]

Michael Faraday (1791–1867)

The poet Coleridge said he went to hear Davy "to increase his stock of metaphors" and asserted that "had [he] not been the first chemist, he would have been the first poet of his age."[6] Toward the end of 1813, with the permission of the Emperor Napoleon, Davy set out on a two-year continental tour, accompanied, as "assistant in experiments and writing," by Michael Faraday, recently engaged as assistant at the Royal Institution.

Born in 1791, son of a blacksmith, Michael Faraday was apprenticed at age thirteen to a bookseller. Self-taught in the basics and in chemistry, he began to attend science lectures at the Royal Institution in 1810. His apprenticeship completed in 1812, he went to work as a journeyman bookbinder. He tells what happened next:

When I was a bookseller's apprentice I was very fond of experiment and very adverse to trade. It happened that a gentleman [a customer of his master and] a member of the Royal Institution, took me to hear some of Sir H. Davy's last lectures in Albermarle Street. I took notes and afterwards wrote them out more fairly in a quarto volume. My desire to escape from trade, which I thought vicious and selfish, and to enter into the service of Science, which I imagined made its pursuers amiable and liberal, induced me at last to take the bold and simple step of writing to Sir H. Davy, expressing my wishes . . . at the same time I sent the notes I had taken of his lectures.[7]

According to Davy's own account, the notes must have made an impression. He recorded this exchange with an acquaintance:

"What am I to do? Here's a letter from a young man named Faraday; he has been attending my lectures and wants me to give him employment at the Royal Institution—what can I do?"

"Do? Put him to work washing bottles; if he is good for anything he will do it directly, if not, he will refuse."

"No, no, we must try him with something better than that."[8]

So Faraday escaped the role of "bottle washer" and, in March of 1813, became assistant to Davy, leaving with him some months later on their continental junket. Faraday published his first research paper in 1816; by 1823 he was a fellow of the Royal Society. The six Christmas lectures for young people at the Royal Institution, initiated in 1826, were given on nineteen occasions by Faraday. (The tradition continues to this day, interrupted only by World War II, but now it is carried to vastly larger audiences by the television services of the British Broadcasting Corporation.) Faraday remained at the Royal Institution for fifty-four years.

Strangely enough, the ground had already been prepared by Isaac Newton, playing an unfamiliar role. The well-known Newton is the one who insisted that a description of the phenomena must precede speculative inquiry about causes.

> . . . to derive two or three general Principles of Motion from Phaenomena, and afterwards to tell us how the Properties and Actions of all corporeal Things follow from those manifest Principles, would be a very great step in Philosophy, though the Causes of those Principles were not yet discover'd.[9]

This is the Newton of the motto *Hypotheses non fingo* [I feign no hypotheses].[10] The other, less well known, speculative Newton is the one who wrote the following to the classics scholar Richard Bentley:

> That gravity should be inate, inherent, and essential to matter so that one body may act upon another at a distance through a vacuum and without the mediation of anything else . . . is to me so great an absurdity that I believe that no man who has in philosophical matters a competent faculty of thinking can ever fall into it.[11]

Having been influenced by this passage (Newton's letters to Bentley were published in 1756), Faraday was predisposed to seek an understanding of the magnet's influence in some intervening medium. The way that iron filings, sprinkled on a card held near a magnet, arranged themselves in curved lines supplied him with a visual manifestation of just such a physical influence permeating the space around the magnet. This led him to replace the notion of forces acting across finite distances with that of lines of force filling all of space. The seed was thus planted in the field that Maxwell would reap.

Maxwell's electromagnetic investigations began in his undergraduate days at Cambridge, where he

> . . . resolved to read no mathematics on the subject till I had first read through Faraday's *Experimental Researches in Electricity*. I was aware that there was supposed to be a difference between Faraday's way of conceiving phenomena and that of the mathematicians. . . . For instance, Faraday, in his mind's eye, saw lines of force traversing all space where the

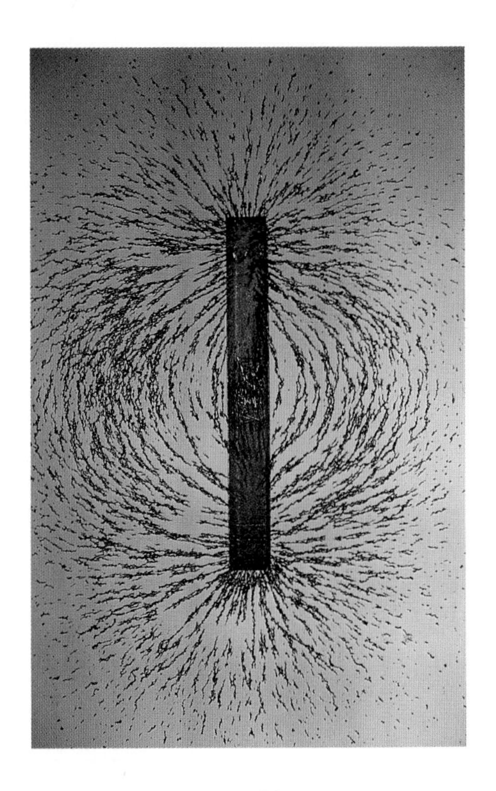

A bar magnet and iron filings

mathematicians saw centres of force attracting at a distance: Faraday saw a medium where they saw nothing but distance . . . . As I proceeded with the study of Faraday, I perceived that his method . . . [was] capable of being expressed in ordinary mathematical forms.[12]

Maxwell's first paper on this subject (1855), written largely in that Cambridge period, is called "On Faraday's Lines of Force." In this, and in several subsequent papers, Maxwell made speculative analogies with fluid motion. Ten years later, with the publication of "A Dynamical Theory of the Electromagnetic Field," the analogies disappeared; this definitive paper is firmly grounded in experiment and in general dynamical principles. Like Newton, Maxwell succeeded by insisting on an economical description of the phenomena.

Following Faraday, Maxwell used the term "field" to describe the physical state of affairs in a region of space that can manifest a certain kind of force: the *gravitational* field of the Earth; the *electric* field of a charge; the *magnetic* field of an electric current. These terms also relate the various kinds of fields to their sources. In essence, Maxwell extended Faraday's induction discovery—that, as shown by the flow of electric charge, a changing *magnetic* field creates an *electric* field—to the reciprocal statement, that a changing *electric* field generates a *magnetic* field. Here is the unification of electric and magnetic fields in the *electromagnetic* field.

Suppose that at a point out in otherwise empty space—a vacuum—a magnetic field changes. (It changes because, somewhere else, at an earlier time, an electric current changed; but that is *not* the significant point here.) The changing magnetic field creates a changing electric field, which in turn regenerates a magnetic field. That is, as *time* elapses, at that point there is an *oscillation* between the two kinds of fields. In addition, the fields vary from one point of *space* to another. All this is reminiscent of a more familiar phenomenon, the motion of waves. For example, if you are floating in a boat on the ocean surface, the passage of ocean waves by that point in space is experienced in the course of time as a rhythmic up and down motion of the boat—it oscillates (see next page). And, at a given instant, if you look out, there in the distance—removed in space—are the successive troughs and crests of the advancing waves. In short, Maxwell's unification of electricity and magnetism led to his prediction of *electromagnetic waves.*

Ocean waves passing a boat

Electrical Units

The speed of electromagnetic waves could be calculated from known quantities, the electrostatic and electromagnetic *units* of electric charge. The electrostatic unit is defined by the electric force between charges at rest; the electromagnetic unit, by the magnetic force between charges in motion. In the very origin of the word electric (see Box 1·4), we have a reminder that static electricity was familiar to the ancients; electromagnetic forces associated with motion, on the other hand, are of recent discovery, because they are comparatively weak at ordinary speeds. Accordingly, charges moving at some such standard speed, say one meter per second, must be considerably larger than stationary charges if the magnetic force is to equal the electric force: the ratio of electromagnetic and electrostatic units of charge is a big number. To put it another way, that ratio is the speed at which the *same* charges will exert electric and magnetic forces of the *same* strength. And that is just the speed that Maxwell predicted for his electromagnetic waves.

Although the ratio of units was not known very well in Maxwell's time, its accuracy was good enough to make it unmistakably clear that Maxwell's electromagnetic waves travel at the speed of light. Maxwell himself embarked on an extended experimental

BOX 1·4 ***Amber***

The yellow resinous substance, amber, which is found on the shores of the Baltic Sea, was an object of commerce even in prehistoric days. There is a tradition that Thales of Miletus, one of the founding fathers of Greek science in the sixth century B.C. was familiar with a strange property of amber: when rubbed, it acquired the power of attracting bits of leaves or feathers. The Latin name for amber is *electrum* (the related Greek word survives in the name of the first discovered negatively charged particle, the electron). In 1600, William Gilbert (1540–1603), personal physician to Queen Elizabeth, published the first scientific study of magnetism and electricity, *On the Magnet.* He introduced the name "electrica" for substances that, like amber, attract light bodies after being rubbed. John Dryden (1651–1700) may have overestimated the durability of temporal power when he wrote:

Gilbert shall live till loadstones cease to draw
Or British fleets the boundless ocean awe.

(The loadstone is a magnetic iron ore, magnetite, when it possesses polarity. It was the original compass.)

program to measure the ratio of units with greater precision. By the end of the nineteenth century, the averaged results of various measurements, by a number of investigators, of the ratio of electrical units, and the averaged measurements of light speed, agreed with each other better than did the various individual measurements of a particular kind. Maxwell himself said this about the speed ("velocity") that is the ratio of charge units:

This velocity is so nearly that of light, that it seems we have strong reason to conclude that light itself (including radiant heat, and other radiations if any) is an electromagnetic disturbance in the form of waves propagated through the electromagnetic field according to electromagnetic laws.[13]

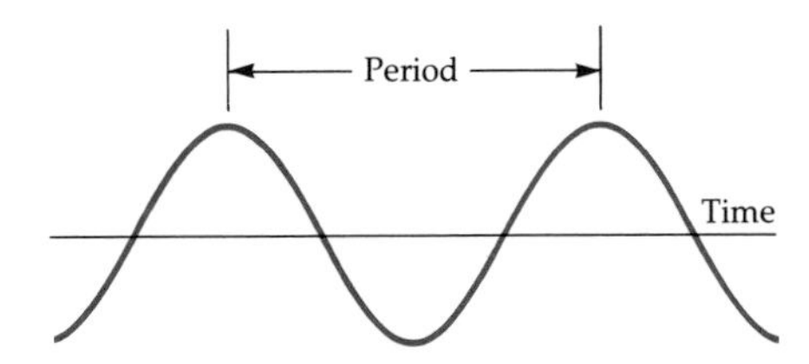

Frequency
A wave of definite frequency is an oscillation that recurs after a definite temporal interval called the period. The frequency, or rate of oscillation, is the ratio between any given number of oscillations and the time that they require. There is one oscillation in a period. Therefore, the frequency is the inverse of the period.

Frequency

When Maxwell speaks separately of light, radiant heat, and other radiations (if any), what distinguishing characteristic does he have

Examples of radio devices

Left: Large antenna array sensitive to long wavelengths.
Middle: Multiuser microwave tower.
Right: Radio telescopes.

in mind? Consider the harpist, plucking strings and thereby setting them into vibration, each with its characteristic rate of vibration, or *frequency* (see preceding page). A string vibrating at a certain frequency sets the air into oscillation at the same frequency. The resulting sound waves reach our ears, where they are recognized as tones. Tone is the physiological counterpart of acoustical frequency. However our ears are indifferent to vibrations that are pitched lower than about 30 per second or higher than about 30,000 per second, a range of a factor of 1,000.

Similarly, for light, *color* is the physiological counterpart of frequency. But our eyes give us a much narrower window on the world than do our ears. From the lowest frequency of red light to the highest frequency of violet light is about a factor of 2. Frequencies just below red (*infrared*) are what Maxwell meant by radiant heat. When we turn on an electric stove, heat is felt at a distance before the coils begin to glow red. The frequencies just beyond the other end of the visible range are the ultraviolet rays that, for example, produce the tanning effect of sunlight.

Now, what about "other radiations"? Maxwell's theory posed a challenge—to create and detect, by purely electrical means, waves of hitherto unknown frequencies. More than twenty years would pass before the German physicist Heinrich Rudolph Hertz (1857–1894) would succeed in this and give the world the first ex-

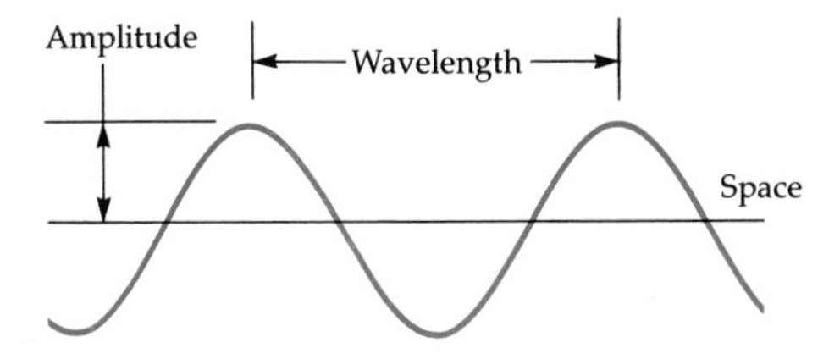

Wavelength

A wave of definite wavelength is an oscillation that recurs after a definite spatial interval, which is the wavelength.

ample of radio waves.* It had been known for some time that an electric circuit, containing a condenser (in which electric fields are concentrated) and a coil of wire (where magnetic fields are localized), oscillates at a definite frequency. The problem was how to detect the electromagnetic waves that it presumably radiated. Hertz solved this by using a wire circle with a tiny gap, across which sparks would signal the presence of electromagnetic waves. Such was the primitive beginning that, in less than a hundred years, would give rise to the installation of thousands of radio and television stations, the microwave transmission of messages between points on Earth and from Earth to satellites in Earth orbit, interplanetary communication, radio telescopes—the list is endless.

Wavelength

The term "microwave," most familiar today in the context of ovens, reminds us that waves are characterized not only by frequency, but also by *wavelength,* the distance between successive crests. The range of known electromagnetic wavelengths is enormous. Wavelengths of many kilometers have been produced. Ordinary (A)mplitude (M)odulation radio uses wavelengths of hundreds of meters, whereas (F)requency (M)odulation radio operates at wavelengths of several meters. Public television station KCET in Los Angeles, for example, broadcasts at a wavelength of slightly more than half a meter. One speaks of *microwaves* when the wavelength is on the scale of centimeters. In contrast, the wavelengths of *visible light* range from about 80-millionths to 40-millionths of a centimeter. And the *X-rays* discovered by Wilhelm Conrad Röntgen (1845–1923) in 1895 are no longer mysterious; they are electromagnetic radiations with wavelengths some thousand times as small as those of visible light. Even shorter wavelengths are

Electromagnetic spectrum

To the left of zero, wavelengths are given in powers of ten (1 = 10 cm, 2 = 100 cm, etc.); to the right of zero, which is 1 cm, wavelengths are given in inverse powers of ten (1 = $\frac{1}{10}$ cm, 2 = $\frac{1}{100}$ cm, etc.).

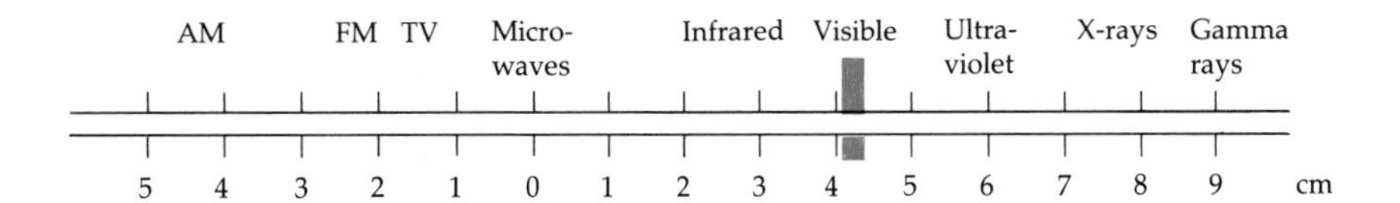

*One could wish that Maxwell had lived to see the confirmation of his theory by Hertz. But nine years before that happened, when at the peak of his mature powers, he was struck down by the same disease and at the same age, forty-eight, as his mother before him.

those of the *gamma rays*, emitted by the nuclei of atoms, first identified as one of the three types of radiation from radium, the unstable element discovered in 1898 by Marie and Pierre Curie.

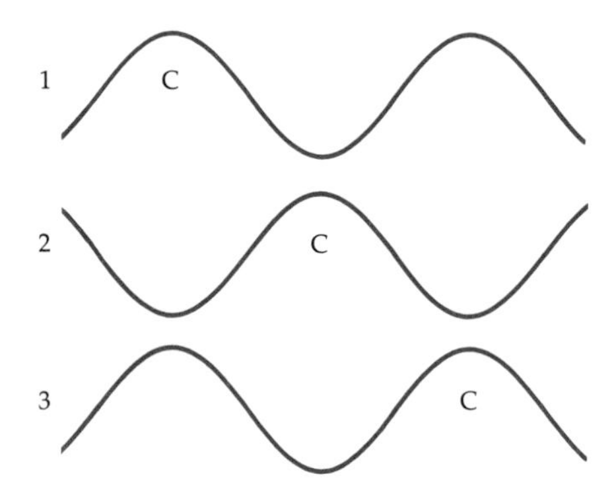

Wave speed

A wave moving to the right: (1) at a certain time; (2) half a period later; (3) a full period later. In an interval of one period, crest C has traveled one wavelength. Therefore, the speed of the wave is the ratio of the wavelength to the period; this is also the product of the wavelength and the frequency.

Speed

Wavelength and frequency are alternative, equivalent ways in which to specify a particular type of wave. How are they related? Wavelength (the distance between crests) and frequency (the number of oscillations in a unit of time) are, respectively, spatial and temporal measures of the wave. They are connected by the measure of the rate at which space is covered in time, or the *speed* of the wave. The *frequency*, the number of waves that pass a point in a unit of time, *multiplied* by the length of each wave (the *wavelength*) gives the total distance that the succession of waves has traveled in a time unit, which is its speed. The numerical value of the wave speed depends on the choice of units for space and time. Thus, for light, it is about 300,000 kilometers per second (km/s), or about 0.3 kilometer per microsecond (km/μs), a microsecond being a millionth of a second. The latter way of putting it tells us directly that a radio wave of 300-meter (0.3 km) wavelength, for example, has a frequency of a million vibrations per second, or one megaHertz (MHz), a unit that honors the first person to detect radio waves.

Hertz's response to the challenge of Maxwell's theory is not the whole story. It must also be possible, by purely electrical means, to produce *known* radiations, especially visible light. This

Electron synchrotron

High energy electrons are constrained by magnetic fields to move in a nearly circular path.

challenge was first met in 1947. The instrument required is an *electron accelerator,* the synchrotron, in which electrically charged particles that swing in circles under the guidance of a magnetic field—a changing electric current—radiate strongly into a range of frequencies that is controlled by the energy of the electrons. Present-day devices can produce any part of the electromagnetic spectrum—the whole range of frequencies and wavelengths—extending to the far X-ray region. More will be said about this matter in Chapter 3.

THE CONFLICT

Both Newton and Maxwell, separated in time by two centuries, created theories that organized vast areas of experience, one in the realm of mechanical motion at moderate speeds, the other for wave phenomena propagated at the highest known speed. Both theories continue to be valuable, supplying understanding and control of their respective types of phenomena. But are the two theories compatible, or might there be a conflict between them?

Looking back at it, the inevitability of a conflict could have been foreseen. In Newtonian mechanics, time is absolute,* which means that two observers, no matter how far apart they may be, and no matter how rapidly they may be moving relative to each other, agree completely on the time assigned to any event. Implicit in this notion is the assumption that the two observers can exchange signals at unlimited speed—instantaneously. But, to the best of our knowledge, there is no physical agency that travels faster than light, which, according to Maxwell's theory, has a definite, invariable speed. (Under ordinary circumstances, the speed of light is so great that light itself appears to be the instantaneous signaling mechanism.) If no instantaneous signaling mechanism exists in nature, there is something shaky about the foundations of Newtonian mechanics.

If that is so, it must be possible to find an experimental inconsistency, at least in principle. In such an experiment, bodies would have to be moving so rapidly that their speeds would *not* be negligible compared with that of light; the speed of light could then no

---

*As is space. Newton's great contemporary, Gottfried Wilhelm von Leibnitz (1646–1716), who was elected a foreign member of the Royal Society in 1673, disagreed. To him, space was an aspect of phenomena; space, and time, were relative concepts.

The teen-aged Einstein

longer be regarded as infinitely large. But that kind of speed is not easily come by in ordinary circumstances. The possibility of a confrontation between Newton and Maxwell must have seemed remote at the end of the nineteenth century.

Yet, although it remained unknown to the world at large, such a confrontation *was* realized in an insight of a sixteen-year-old high-school dropout who was regarded by his elders as backward and indifferent. This was Albert Einstein. He was born in Ulm, Germany, in 1879, the year of Maxwell's death, and grew up in Munich. His education began at a Catholic elementary school, where the strict discipline and mechanical teaching repelled him. When he was ten years old, he entered the Luitpold Gymnasium (roughly equivalent to a high school), but this was no improvement. Like Maxwell before him, he was *different*.

His intellect had first been stirred at the age of five by the gift of a magnetic compass. Then, at age twelve, he encountered and was deeply moved by the wonders of Euclidean geometry. An uncle had told him about the theorem of Pythagoras, relating the sides of a right-angle triangle. About this he later said, "After much effort I succeeded in 'proving' this theorem on the basis of the similarity of triangles." Then, in a small geometry book that came into his hands, Einstein saw, as he put it, "assertions which—though by no means evident—could nevertheless be proved with such certainty that any doubt appeared to be out of the question. This lucidity and certainty made an indescribable impression upon me."[14] Such was the student who, at age fifteen, was advised to leave the Gymnasium without a diploma because he would never amount to anything and because his indifference was demoralizing. Self-taught in mathematics and physics, Einstein entered the Polytechnical Institute in Zürich, Switzerland, at the age of seventeen.

But by then he had already had a profound insight. It was, in his own words:

> . . . a paradox upon which I had already hit at the age of 16. If I pursue a beam of light with the velocity c (velocity of light in a vacuum), I should observe such a beam of light as a spatially oscillatory electromagnetic field at rest. However, there seems to be no such thing, whether on the basis of experience or according to Maxwell's equations.[15]

Here was Einstein's first use of a favorite device—the *thought experiment*. Through the power of imagination, practical constraints can

BOX 1·5 ***Similar Triangles and the Theorem of Pythagoras***

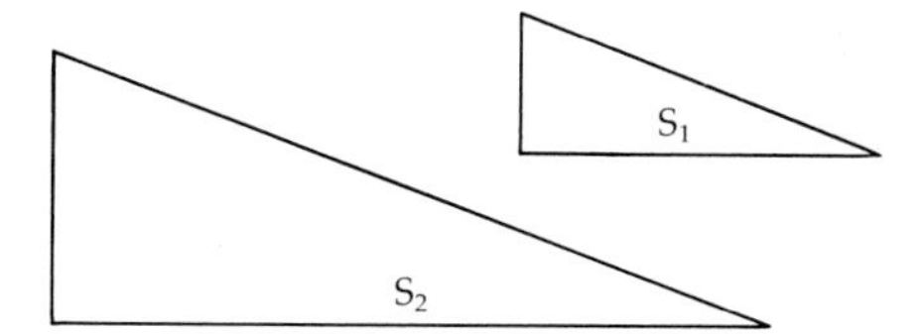

Triangles S_1 and S_2 have the same angles and differ only in scale; each side of the larger triangle is twice the corresponding side of the smaller triangle. They are an example of similar triangles.

The twelve-year-old Einstein may have proved Pythagoras' theorem "on the basis of similar triangles" in the following way.

Triangles 1 and 2, at the left below, and the triangle comprising both of them have the same angles;* they are similar triangles. The corresponding ratios of any two sides will be the same for all three triangles. Therefore

$$\frac{a}{c} = \frac{c_1}{a} \quad or \quad a^2 = c_1c$$

and

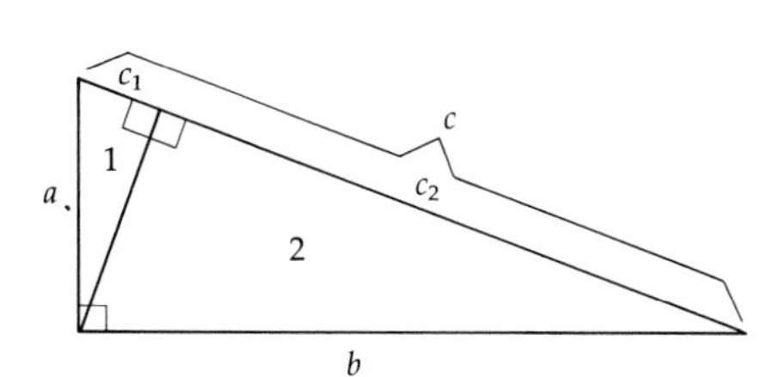

$$\frac{b}{c} = \frac{c_2}{b} \quad or \quad b^2 = c_2c$$

$$Add: a^2 + b^2 = (c_1 + c_2)c$$

But $c_1 + c_2 = c$. Therefore,

$$a^2 + b^2 = c^2.$$

On the other hand, he may have chanced on a more-intuitive proof that uses the areas of similar triangles. Inasmuch as a triangle is a two-dimensional figure, the area of the larger of the triangles S_1 and S_2, in which all sides are doubled, is $2 \times 2 = 2^2$ times that of the smaller triangle. More generally, the ratio of the areas of two similar triangles is the square of the ratio of the lengths of any corresponding side.

Among all similar right-angle triangles, the area of any particular one is proportional to the squared length of the slant side of that triangle. Now, the large triangle (with slant side c) has an area that is the *sum* of the areas of triangle 1 (with slant side a) and triangle 2 (with slant side b). When the common constant of proportionality is removed from the area relation of the three similar right-angle triangles, we have just

$$a^2 + b^2 = c^2.$$

*Each triangle has a 90-degree angle; both smaller triangles share an angle with the large triangle; the sum of the three angles of any triangle equals 180 degrees.

be cast aside and the implications of theories examined at their extreme limits.

In Newtonian mechanics, motion at any speed is possible, and any uniformly moving object can be caught up with by another body, after which the relative speed of the two is zero. Familiar examples are a police car overtaking a miscreant driver on a freeway or a bicycle on the bank of a river catching up with a slowly moving boat. Einstein applied this Newtonian reasoning to light. In principle, then, it should be possible for an observer to catch up with a beam of light and then travel at the speed c. To that observer, the light wave will be seen as oscillating in space but *not* moving. This, however, is a situation that, according to Maxwell's theory, can never happen, for the speed of light is always c, the ratio of the electromagnetic and electrostatic units of charge.

That is the confrontation between Newton and Maxwell. Which of the two theories is wrong? What is meant here by "wrong" is that, although the theory still works admirably in its original domain, it fails when pushed far outside that domain. It was suggested earlier that Newtonian mechanics is on shaky ground because of Maxwell's prediction that the speed of light is finite and invariable. But how reliable is *Maxwell's* prediction?

Perhaps the speed of the light emitted from a moving object depends on the speed of that object, just as a bullet fired straight ahead from a moving car has its speed increased by the speed of the car, whereas one fired backward has its speed decreased by the same amount. That result would be expected in Newton's corpuscular theory of light. But does light behave in this way? If so, the foundations of Newtonian mechanics might yet be secure. The only way to settle the question is by experiment. And in 1913 the Dutch astronomer Willem de Sitter (1872–1934) pointed out that Nature had already supplied the material for such an experiment in the form of double, or binary, stars.*

Binary Stars

The first double star was discovered in 1650, less than a half-century after Galileo turned his telescope on the sky. The Italian as-

---

*The reader is alerted that, for clarity of exposition, the historical order of development is not always respected. Einstein's first publication on relativity, for example, could not make use of de Sitter's observation; it would not appear for another eight years.

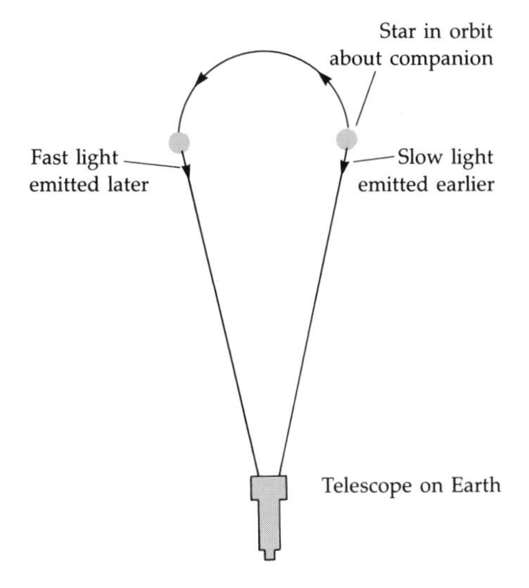

DeSitter's argument

If the speed of light depended on the motion of its source, in the Newtonian manner, slower light emitted earlier and faster light emitted later could arrive at the same time.

tronomer Jean Baptiste Riccioli (1598–1671) observed that the star Mizar, in the middle of the handle of the Big Dipper, appeared through his telescope as *two* stars. In the next century and a half, many other closely separated pairs of stars were discovered telescopically. Between 1782 and 1821, the British astronomer Sir William Herschel (1738–1822) published three catalogs that listed a total of more than eight hundred double stars. In the first publication, Herschel suggested that among these double stars might be pairs revolving about each other under the action of Newtonian gravitation. Many years later, he announced that some of them *had* changed their relative positions, as this view required.

De Sitter remarked that a number of such pairs of stars have orbits that are nearly edge-on to our line of sight, so that each star is alternately approaching and receding as it revolves about its companion. Now, if the speed of the light emitted from an orbiting star depended on the speed of that star, as in the Newtonian manner, light coming to us from any part of the orbit where the star is approaching would be speeded up, and light coming from any part of the orbit where the star is moving away would be slowed down. It is then possible that the speed of the star and its distance from the Earth would be such that the light emitted as the star approaches us would arrive at the same time as the slower-moving light emitted earlier, when the star was moving away. We could even see the star on opposite sides of its orbit at the same time. That is a very special situation however; in general, there would be multiple images and spread-out images if the speed of light really did depend on the speed of its source. But, in fact, we always see double stars moving in well-behaved elliptical orbits about each other.

So Maxwell was right, at least in that the speed of light is not affected by the motion of the light source, as Newton would have had it. To return to Einstein's thought experiment, imagine that astronauts in a powerful rocket attempt to catch up with a beam of light emitted in pulses from a lamp on the Earth. After the rocket has reached a respectable speed, the astronauts look back and see the Earth and lamp rapidly receding in the distance. To them, that lamp is a moving source of light. But the observations of double stars have told us that the motion of the source has no effect whatever on the speed of light. And so the astronauts observe the light moving by them at the same speed c, despite their efforts to catch up with it. The mission is a failure.

Galileo Galilei (1564–1642)

Relativity

But wait! Why do we assume that the great speed at which the astronauts are moving is unimportant? We know that motion of a light source relative to an observer has no effect on the light speed, but is that also true for a motion of the observer relative to the light source? According to the Newtonian laws of mechanics applied to moving bodies, two different situations with the same relative motion are indeed equivalent; that is the *principle of relativity*, first recognized by Galileo, who described it this way:

> Shut yourself up . . . in the main cabin below decks on some large ship . . . . Have a large bowl of water with some fish in it. . . . With the ship standing still . . . the fish swim indifferently in all directions. When you have observed . . . carefully . . . have the ship proceed with any speed you like, so long as the motion is uniform. . . . You will discover not the least change . . . nor could you tell . . . whether the ship was moving or standing still.[16]

Does this principle also apply to the wave motion that is light? First let us look at a more familiar type of wave.

Water waves can be generated by the movement of a boat on the surface or by dropping a rock into the water. But whether or not the source has a motion parallel to the surface, all the waves, as observed by someone at rest in the water, roll away with the same characteristic speed. An observer in motion, however, sees a changed speed. If he moves toward the source of the waves, they arrive more rapidly; should he move in the same direction as the waves, they appear to travel more slowly. It is also possible to catch up, to move with the waves, as when a surfer balances precariously near the crest of a wave. And there is another possibility: let the wind spring up, driving the water before it. Then, waves moving in the direction of the wind are speeded up, whereas those moving against the wind are slowed down.

As these examples show, there is more to consider than simply the relative motion of the source and the observer: the medium in which the waves travel also plays a role. It is still true that only relative motion counts, but it is the relative motion of source, observer, and *medium*. For most of the nineteenth century, the grip of the mechanistic Newtonian viewpoint was so strong that virtually no one disputed that light waves must also travel in a medium. If

there are waves, which is to say, undulations, then surely *something* must be undulating. It was called the *ether.*

Ether

Maxwell had this to say about the historical proclivity for inventing "aethers":

Aethers were invented for the planets to swim in, to constitute electrical atmospheres and magnetic effluvia, to convey sensations from one part of our bodies to another, and so on, till all space had been filled three or four times over with aethers. . . . The only aether which has survived is that which was invented by Huygens to explain the propagation of light. [It was reintroduced at the beginning of the nineteenth century in the revival of the wave theory of light by Thomas Young, under the name luminiferous ether.] . . . The properties of this medium . . . have been found to be precisely those required to explain electromagnetic phenomena.[17]

Strange and wonderful were those properties. The ether had to be dense enough and elastic enough to support electromagnetic oscillations of any frequency, but offer no resistance to the passage of matter through it.* Maxwell himself seemed ambivalent about the reality of the ether. He gave it lip service, and yet he called it a "most conjectural scientific hypothesis." His description of light as " . . . waves propagated through *the electromagnetic field*" (my italics) is thoroughly modern. And it was he who suggested the definitive experiment to test the ether hypothesis, the underlying ideas for which are sketched below.

If the Earth, in its orbit about the Sun, moves through the ethereal medium, does it drag the ether with it, or, in a phrase of Thomas Young, does it let the ether slip through "like the wind through a grove of trees"? The observed *aberration of starlight* (see Box 1·6), the apparent change in direction of a star as the Earth moves from one side of the Sun to the other, shows that light travels in a straight line from a star to the moving Earth; therefore, the ether in which the light travels is *not* pulled along with the Earth. On the Earth, in motion through the ethereal medium, the relatively moving ether will indeed seem "like the wind," and, just as moving air that carries sound changes the speed of the sound, that ether wind carrying light should change the speed of the light along the direction of the wind, which is set by the direction of the Earth's motion in the ether. This change might be detected, Maxwell realized, by sending two light beams on different paths of *equal length*, because the change in the *speed* of light moving in different directions would produce *unequal travel times*. But to mea-

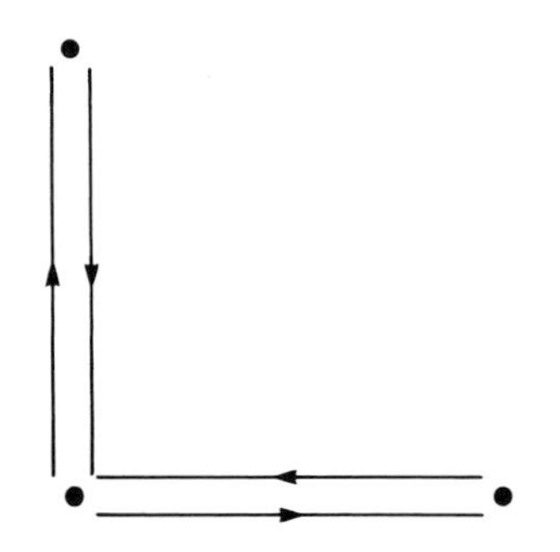

Maxwell's idea

The two paths, at right angles, are of equal length. If the two round trips are conducted at equal speed, they will take equal time. But, if a wind (ether or atmospheric) is blowing along one of the directions, the travel time (for light or airplanes) will be different.

*Another major difficulty was this. The waves that travel through ordinary solid bodies are of two types: longitudinal—the vibration is along the direction of motion (these are sound waves); and transverse—the vibration is perpendicular to the direction of motion. As a discovery of Christian Huygens first indicated, however, light waves are entirely transverse. (This has become quite familiar through the use of polarizing glasses, which screen out one of the two possible transverse vibrations. Two such polarizers, held at right angles, stop light completely.) To explain the absence of longitudinal light waves, it was necessary to assume that the ether behaved like an *incompressible* solid.

BOX 1·6 ***Aberration of Starlight***

In the seventeenth century and the first quarter of the eighteenth, the hunt went on for a direct proof of the Copernican theory of the solar system. If the Earth revolves about the Sun in a year's time, the direction of a certain star should be different after six months, when Earth is on the opposite side of its orbit. This effect is familiar to us in the shift of a nearby object when we look at it first with one eye closed and then the other; it is the basis of stereoscopic vision.

By 1725, James Bradley (1693–1762) observed such a shift of a star that happens to pass directly overhead at London. But something was amiss; the maximum shifts occurred at the wrong times of the year, when the Earth was displaced from the expected positions by a quarter of its orbit. Two years later, while sailing on the Thames in a steady breeze, Bradley became fascinated by a flag atop the mast, which was changing direction in response to alterations in the boat's course. That gave him the vital clue: The effect he had observed for starlight was produced, not by the Earth's changing *position* in the orbit, but by Earth's changing *motion* in its orbit.

One favorite analogy for this effect of motion imagines someone standing in vertically descending rain, using a vertically held umbrella for maximum shielding. When that person walks, the umbrella must be inclined forward to maintain the shielding, the more so the faster the pace.* Turn the rain into starlight, the umbrella into a telescope, and the umbrella carrier into the Earth in its orbit, and you have Bradley's explanation of the aberration of starlight.

This picture, of light moving in a straight line from a star to the moving Earth, gives a complete account of the observed changes in direction of the star in the course of a year. But, if light moved in an ethereal medium that was dragged along with the moving Earth, no such effect would exist. To use the rain analogy, it is as though, when the person begins to walk, a wind springs up, driving the rain in the same direction, and with the same speed, as the person moves.

*An equivalent effect appears when the person remains stationary and a wind springs up to drive the rain horizontally. The discussion of the Michelson-Morley experiment (see text) involves something similar.

sure that time difference would require a precision of greater than one part in 200 million! (More about this later.) Maxwell concluded that the experiment was impossible.

In 1879, the last year of his life, Maxwell wrote to the astronomer David Todd (1855–1939), at the Nautical Almanac Office in Washington, D.C., asking whether the data on the eclipses of Jupiter's moons were accurate enough to detect the Earth's motion through the ether. In that letter he comments on the "impossibility" of earthbound optical experiments for this purpose. The letter came to the attention of a colleague of Todd, Albert Michelson (1852–1932), who had already carried out the best measurement, as of that time, of the speed of light in air.

An Analogy

Michelson rose to the challenge of doing the "impossible" optical experiment. But before we come to the key idea that brought such fantastic precision within reach, let us use an analogy with a more familiar situation to set out the general plan of the experiment, as Maxwell first conceived it. Let the speed of light in the ether be replaced by a fixed speed of airplanes in still air, and let the ether wind that changes the speed of light be replaced by the high atmospheric wind (the jet stream) that significantly alters the ground speed of airplanes. For the purposes of this story, assume that the jet stream flows at a constant speed in a known direction.

Problem: Find the speed of the jet stream by timing airplanes sent on round trips in different directions, but of equal length, that are flown at a fixed airspeed.

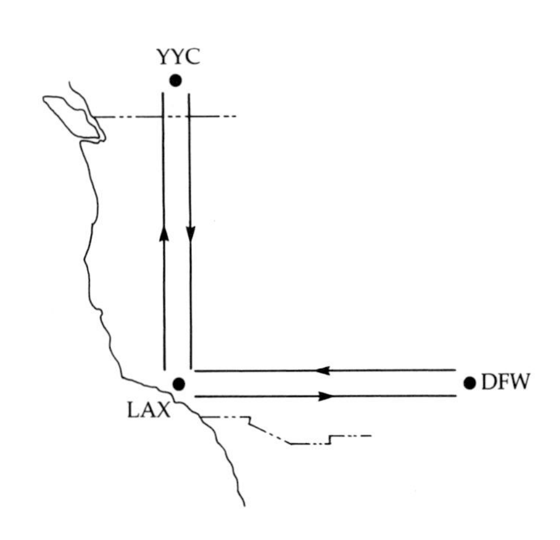

A dispatcher at Los Angeles International Airport (LAX) responds this way. He sends off two airplanes at the same time, one eastward to Dallas (DFW), and the other northward to Calgary (YYC). These two cities are almost equally distant from LAX. The pilots are instructed to fly to their respective destinations and return to LAX, all the while steadily maintaining a common airspeed. By noting which plane arrives first and clocking its lead, the dispatcher verifies that the jet stream is running and finds its speed.

To see how this works, let us suppose that each airplane has an airspeed of exactly 1,000 km/h and that the jet stream is blowing due east at 600 km/h. Now consider the eastbound pilot. Relative to the ground, his 1,000 km/h is increased by the 600 km/h speed

Eastbound 5 → 3 → = 8 →

Westbound ← 5 = ← 2

3 →

West-to-east and east-to-west trip speeds

Speed in units of 200 km/h.

of the jet stream, and so his ground speed from LAX to DFW is 1,600 km/h. He arrives at DFW after a certain period of time that we call *one* time unit. On his return, however, he is fighting the jet stream, and his ground speed is 1,000 less 600, or only 400 km/h, one-fourth his speed from LAX to DFW. His return journey, therefore, takes four times as long, or *four* time units. So the round trip requires 4 plus 1, or 5, time units.

Now, how about the plane to Calgary? In still air, it travels at 1,000 km/h; however, the jet stream is blowing the plane off course, to the east, at 600 km/h. To fly due north, the pilot must head into the wind. The airspeed of 1,000 = 5×200 km/h is directed to the northwest to such an extent that its combination with the easterly directed jet stream speed of 600 = 3×200 km/h produces a northerly directed ground speed of $4 \times 200 = 800$ km/h. (This 3-4-5 right-angle triangle relation is the simplest example of the Pythagorean theorem that so enthralled the twelve-year-old Einstein.) The trip from Calgary back to Los Angeles is similar. Again the pilot must head into the wind, this time to the southwest, and again his ground speed is 800 km/h. In other words, our LAX-to-YYC pilot has a ground speed in both directions that is exactly half that of the plane from LAX to DFW, and so he will require *two* time units each way, or four time units for the round trip.

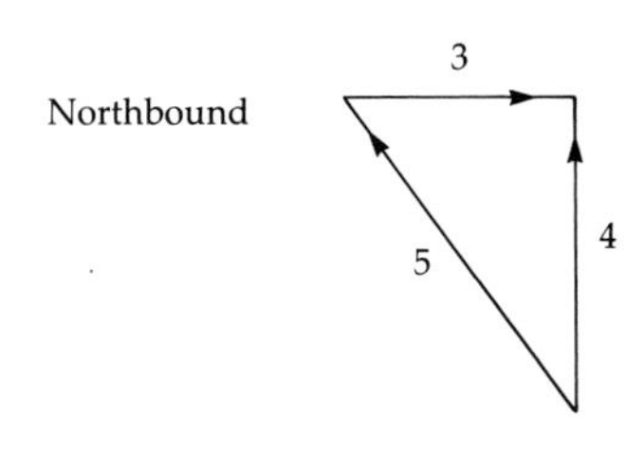

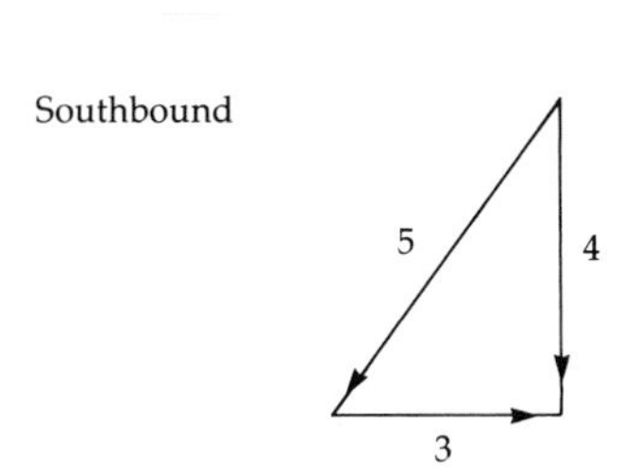

South-to-north and north-to-south trip speeds

Speed in units of 200 km/h. The diagrams also illustrate how rain, dropping vertically but swept by a horizontal wind, falls at a slant.

Now we see the outcome. The round trip perpendicular to the jet stream takes only *four* time units, whereas the round trip parallel to the jet stream takes *five* time units. Our airport dispatcher learns that the jet stream is blowing, and, from the ratio of travel times, 4:5, he works back to the ratio of jet-stream speed to plane airspeed, 3:5.

Remember that all this is an analogy. The two airplanes traveling at a fixed airspeed represent the two light beams traveling at speed c in the ether. The jet stream that changes the ground speed of the airplanes stands for the ether wind that alters the speed of light, as measured on the Earth, which is moving through the ether. The comparison of the round-trip times in perpendicular directions is an attempt to detect and measure the speed for that motion. We could choose simple numbers for our story: the ratio of jet-stream speed (v) to plane airspeed (c) is $v/c = 3/5$. In Michelson's experiment, however, the ratio of the Earth's speed through the ether (v) to the speed of light (c) is much smaller than one. Then the ratio of the round-trip times, perpendicular to and parallel with the jet stream, respectively, is less than one by a very small amount that is given well enough by $\frac{1}{2}(v/c)^2$. (This approximation is not too bad even for the numbers of the airplane analogy, $v/c = 3/5$.)

The Idea

Now, at last, *what* was the inspired idea that Michelson had for doing this impossible experiment? Let us return to Maxwell's remark that an accuracy of at least one part in 200 million is required. We now recognize in this the value of $\frac{1}{2}(v/c)^2$, with v set equal to the Earth's speed in its solar orbit, 30 km/s, which is one-ten thousandth of c, the speed of light. But instead of focusing on the difference in travel *times* for round trips of the two light beams, think of the difference in *distance* covered when the round trip of the quickest beam is completed. For an apparatus with effective dimensions of, say, 20 meters, the experiment must be able to measure a distance of a hundred thousandth of a centimeter, which is about one-fifth the wavelength of visible light.

That gave Michelson his idea: use the wave nature of light, particularly the phenomenon of interference (see Box 1·1), as the basis for a measurement of distance that is sensitive down to a small fraction of a wavelength. His optical experiment begins with a beam of light that falls on a half-silvered glass plate, transmitting part of the beam and reflecting the other part at right angles (the two airplanes take off). The beams travel equal distances to mirrors that reflect them (the planes reach their respective destinations and turn around), directing them back to the half-silvered plate (the round trips are completed), where they rejoin in a single beam. This beam enters a telescope that is focused on the two coincident images of the light source (the time difference between the two round trips is measured). And what do you see in the telescope? An interference pattern (see Box 1·1).

In Michelson's apparatus, the two superimposed and interfering beams of light have a slight spatial displacement because they originate on opposite sides of the silvering line. The variation in

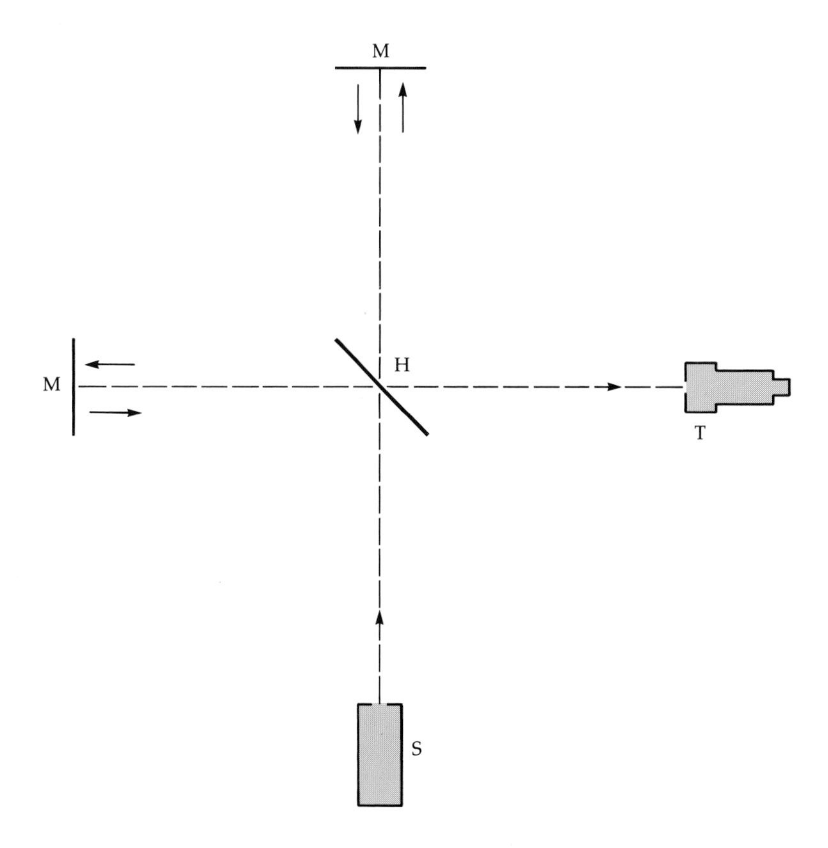

Michelson-Morley experiment

In this diagram, S represents the light source, H the half-silvered glass, M the mirrors (the additional mirrors used for multiple reflections are not shown), and T the telescope.

relative distance of travel for different parts of the image produces interference fringes, which appear as a succession of dark and bright bands of light.

The location of the interference fringes in this pattern depends on the difference in travel time of the two beams. But, to make this difference useful, one must *change* the relative travel time and observe a *shift* of the interference fringes. A change in travel time is equivalent to moving one beam relative to the other. If that displacement were one wavelength, the interference pattern would be shifted by exactly one fringe, and the new situation would be indistinguishable from the original one. Thus, the method is ideally suited to measuring shifts of a fraction of a fringe, displacements that are a fraction of a wavelength.

Very good. But how do you change the relative travel time of the two beams? Back to the airplanes. On another day, the jet stream happens to flow due north, instead of due east. In this situation it is the Calgary-bound plane that returns later than the Dallas plane. In short, to detect the existence of the ether wind, change its direction. In practice, this means changing the orientation of the light paths relative to the ether wind; that is, rotate the apparatus!

The Experiment

With all that behind us, let us return to Michelson and his obsession with the impossible experiment. From 1880 to 1882, Michelson was in Europe, improving his knowledge of physics. It was during his sojourn in Berlin that he first performed his interference experiment, helped by funds from Alexander Graham Bell (1819–1905). His apparatus had several drawbacks. The arms on which the reflecting mirrors were mounted gave a rather short travel distance and were subject to distortion when the instrument was turned. And, even at 2 o'clock in the morning, Berlin traffic set up disturbing vibrations.

The definitive experiment was performed in 1887 when Michelson, then at the Case School of Applied Science in Cleveland, Ohio, joined forces with Edward C. Morley (1838–1923), professor of chemistry at nearby Western Reserve University. (The two schools are currently united as Case Western Reserve.) They introduced improvements to correct the deficiencies of the first apparatus. First, the travel distance was greatly increased, particularly by reflecting the beams back and forth several times. The effective round-trip distance was now about 22 meters, or 40 million wavelengths of yellow sodium light (which was the kind of light used to carefully align the apparatus). Then, to increase the stability of the apparatus and minimize vibrations, they mounted the optical elements on a massive sandstone slab, which floated on a layer of mercury. That also solved the problem of rotating the apparatus. Once set into motion, it would continue for hours, moving so slowly that accurate observations of the fringes could be made at each of sixteen positions marked around a circle.

Finally, observations were made at noon and close to 6:00 P.M. on several consecutive days, beginning July 8, 1887. During this six-hour interval, the rotation of the Earth changed the direction of earthly orbital motion by 90 degrees, relative to the laboratory. The expected maximum displacement of the fringes on rotating the apparatus was four-tenths of a fringe. And what was observed at these two times of the day?

Nothing. No displacement at all. Of course, the experiment had a finite precision; the hundredth part of a fringe might have escaped notice. This experiment has since been repeated, with more modern instruments, to considerably greater accuracy, and always with the same outcome. Yet, even then—to anyone who would see—the null result of the Michelson-Morley experiment carried a clear message: There is no medium—there is no ether.*

Relativity

In describing this historic experiment, I have used the historical reasoning that is tied to the ether concept. The paradoxical aspects of the ethereal medium can be avoided, however, by putting the question that Michelson and Morley answered in the following way: Could it be that Maxwell's theory—in which electromagnetic waves travel at the speed c—refers to some special frame of reference, which is that of an observer who can truly be said to be *at rest?* If so, then an observer who is moving relative to that special frame of reference would measure a speed different from c. The Michelson-Morley experiment proves unequivocally that there is no such special frame of reference. *All* frames of reference in *uniform* relative motion are equivalent, not only for mechanical motions but also for electromagnetic properties. Therefore, a state of absolute rest has no meaning.

But it would be eighteen years before anyone was bold enough to enunciate, as general laws, the two separate statements, emphasized by electromagnetic theory and supported by experiment, that are *irreconcilable* in Newtonian physics. They are the following:

1. *The Principle of Relativity:* The descriptions of *all phenomena* provided by any two inertial observers† in uniform relative motion are equally valid; the laws of physics are the same for both of them.
2. Among those laws is the *absoluteness* of the speed of light in a vacuum; that speed is the same for any two inertial observers in uniform relative motion.

---

*By 1894, it had become possible for Robert Cecil, Lord Salisbury, in addressing the British Association, to admit that "for more than two generations the main, if not the only, function of the word 'aether' has been to furnish a nominative case to the verb 'to undulate.'"[18]

†Light in a vacuum, and material bodies acted on by no forces, move in straight lines for such observers. An observer in *rotation* relative to an inertial observer is *not* an inertial observer. This fundamental distinction is examined in Chapter 6.

The language that Albert Einstein used in 1905 was somewhat more discursive than the preceding statements. Their nearest equivalents in the 1923 English translation[19] of the original paper are

1′. . . . the same laws of electromagnetism and optics will be valid for all frames of reference for which the equations of mechanics hold good.

2′. . . . light is always propagated in empty space with a definite velocity c which is independent of the state of motion of the emitting body.

Notice that the original statement of the second postulate refers to the motion of the emitting body, whereas we have emphasized the observer's motion; the first postulate assures us that the two versions are equivalent.

NOTES

1. *Encyclopaedia Britannica,* 11th ed., s.v. "Royal Society," excerpt from a letter written by John Wallis, 1645.
2. Ibid., s.v. "Newton."
3. *Leonardo Da Vinci* (Reynal and Company), p. 412.
4. **Ivan Tolstoy:** *James Clerk Maxwell* (University of Chicago Press, 1981), p. 136.
5. **R. Siegfried and R. Dott, eds.:** *Humphrey Davy on Geology* (University of Wisconsin Press).
6. *Encyclopaedia Britannica,* 11th ed., s.v. "Davy."
7. **H. Boorse and L. Motz, eds.:** *The World of the Atom* (Basic Books, 1966), p. 317.
8. Ibid.
9. Ibid., p. 103.
10. **Alexandre Koyre:** *Newtonian Studies* (University of Chicago Press, 1968), p. 35. In 1956 this author pointed out that "feign," not "frame," is the correct translation of *fingo.* It is the pejorative sense of "feign hypotheses"—make unsupported speculations—that reconciles Newton's famous dictum with the undoubted fact that he *did* introduce hypotheses, those supported by experimental evidence (Phaenomena).

11. **Boorse and Motz:** *World of the Atom,* p. 319.
12. **James Clerk Maxwell:** *A Treatise on Electricity and Magnetism,* 1873 (Dover).
13. **Boorse and Motz:** *World of the Atom,* p. 343.
14. *Albert Einstein,* Library of Living Philosophers, vol. 7 (1949), p. 9.
15. Ibid., p. 53.
16. **Galileo Galilei:** *Dialogue Concerning the Two Chief World Systems* (University of California Press, 1962), p. 186.
17. *Encyclopaedia Britannica,* 11th ed., s.v. "aether."
18. Ibid., s.v. "light."
19. **A. Einstein, H. A. Lorentz, H. Minkowski, and H. Weyl:** *The Principle of Relativity,* 1923 (Dover).

2

MARKING TIME

ANNUS MIRABILIS

The year 1905 was a miraculous one for science. A totally unknown physicist produced not one but three revolutionary papers in physics that year. He was Albert Einstein.

When we left him at the age of 17, he had entered the Polytechnic Institute in Zürich, Switzerland. Surely, clear sailing was now ahead? Not at all. The courses bored him and disaster loomed, were it not for his friend, Marcel Grossmann, whose classroom notes enabled Einstein to pass the examinations. But Einstein had antagonized his professors, and no academic position was forthcoming on his graduation in 1900. After two years of surviving on odd teaching jobs, and becoming a Swiss citizen, he was again rescued by Marcel Grossmann, who managed to get him a position as junior patent examiner in the Swiss patent office at Bern. Here, financially secure at last, and isolated from the mainstream of physics, Einstein found the time to form the insights that would lead physics into the twentieth century.

The first of the three papers is the one cited in Einstein's Nobel Prize award for 1921. It has the formidable title "Concerning a Heuristic Point of View about the Creation and Transformation of Light." In it, Einstein began by contrasting the two accepted ways of describing the distribution of energy in space. According to the mechanistic, Newtonian attitude, the energy of a body is concentrated in the various *particles* that constitute it, whereas in the electromagnetic, Maxwellian view, the energy is spread throughout the region occupied by the electromagnetic *field.*

Albert Einstein in 1905

Then came his astonishing idea that this strict dichotomy of discrete and continuous—particle and field—might be blurred in the atomic world. Light, which, in view of its wave nature, is certainly described by a *field,* could also exhibit *particle* characteristics. He proposed that

> the energy in a beam of light emanating from a point source is not distributed continuously over larger and larger volumes of space but consists of a finite number of energy quanta localized at points of space, which move without subdividing and which are absorbed and emitted only as units.[1]

The energy carried by each quantum of light, or photon, is proportional to its frequency. That relation between energy and frequency had been introduced five years earlier by the German physicist and Nobel Prize winner Max Planck (1858–1947) to describe the connection between the temperature of a body and the color of the light that it radiates. (A heated body will first glow red, then yellow, then bluish white; the light frequencies rise as the temperature mounts.) Planck, however, had not put forward any such spatial picture of the light energy propagating in the form of a particle. Einstein showed the power of his particle concept by applying it to the photoelectric effect (see Box 2.1).

The second of the 1905 papers also bears a ponderous title: "On the Motion of Particles Suspended in Liquids at Rest Required by the Kinetic Theory of Heat." In 1827, the Scottish botanist Robert Brown (1773–1858) had turned a microscope on pollen grains, about one-two thousandth of a centimeter in length, that were suspended in water. He observed them to be in ceaseless agitation. Some seventy-five years later, Albert Einstein was reinventing the general molecular theory of thermal phenomena, unaware that it had already been constructed by the distinguished American physicist Josiah Willard Gibbs (1839–1903). Concerning this work, Einstein later said:

> My major aim in this was to find facts which would guarantee as much as possible the existence of atoms of definite finite size. In the midst of this I discovered that, according to atomist theory, there would have to be a movement of suspended microscopic particles open to observation, without knowing that observations concerning the Brownian motion were already long familiar.[2]

Josiah Willard Gibbs

BOX 2·1 ***The Photoelectric Effect***

High-frequency ultraviolet light falling on the surface of metals produces an outward flow of electrons. This is the photoelectric effect. It raises two severe difficulties for the idea that the energy of light is distributed continuously throughout the electromagnetic field.

First, if the intensity of the light is very low, some time would be required before an electron in the metal could accumulate enough energy to be ejected from the interior; in fact, the relatively few electrons that are liberated begin emerging as soon as the light beam is turned on. Second, if the light intensity is very high, more energy could be transferred to an electron, which should emerge with greater energy; in fact, the energy of the emitted electrons is independent of the intensity of the light. It is the *frequency* of the light that determines the electron energy.

The concept of the photon explains all this. If a photon can be absorbed only as a unit, then all its energy is available for transfer to an electron. Diminishing the intensity of the light lowers the rate at which electrons are produced, but those electrons will appear with no time lag. With higher beam intensity—more photons—the electron flow is increased, but the energy of the individual electrons is not changed. The energy of each electron is supplied by the energy of a photon, which is determined by the frequency of the light.

The ceaseless dance of the pollen grains bears witness to the reality of atoms (see Box 2.2).

The third paper of 1905 has a short title: "On the Electrodynamics of Moving Bodies." This is the paper in which Einstein set forth his special theory of relativity. The two fundamental assumptions upon which this theory is based were given at the end of Chapter 1. Because Einstein's special theory of relativity is a devastating attack on the Newtonian hypothesis of absolute time, this chapter focuses on *time* and its *measurement*. To begin, let us briefly review the evolution of the concepts and methods of marking time, up to the introduction of Einstein's new insights.

BOX 2·2 ***The Reality of Atoms***

Ideas about atoms and molecules, and the interpretation of heat as the energy of their random motion, had been under development since before Maxwell's day. Nevertheless, at the beginning of the twentieth century, skeptics were still arguing that there was no direct experimental evidence for the reality of molecules. That is what Einstein was looking for. He found a way to get it in the theory that Brownian motion is a visible consequence of the buffeting of a microscopic particle by random collisions with invisible molecules. According to this theory, if such a particle were followed over various time intervals, on the average it should show no net tendency to move in one direction or another. The theory also predicts that the average size of the particle's excursions should grow, in a definite way, with increasing observation time. Once experiment had confirmed this prediction, effective resistence to atomism ceased.

CLOCKS

Any uniformly repeating physical process can be used as a clock because it supplies a standard unit of time against which the durations of other events can be measured. The rhythm of time is thrust upon us from birth; the heart beats, the blood pulses. There is something symbolic in the legend that the seventeen-year-old Galileo used his pulse to time the swinging of a lamp in the cathedral at Pisa and discovered that a pendulum beats the same time, independently of the size of the swing (provided the angle of displacement relative to the vertical is kept small). One can see in this event the beginning of modern science, which could hardly exist until such crude anthropocentric methods of measurement were replaced by objective, controllable instruments, such as the pendulum-regulated mechanical clock. It is said that Galileo himself, shortly before he died, suggested this application of his discovery.

There was an important development in 1673, when the remarkable Dutch scientist Christian Huygens (who first saw that Saturn is ringed and who advanced the wave theory of light) realized how to remove the restriction on the displacement angle of the pendulum. The bob of a pendulum, at the end of a rigid rod,

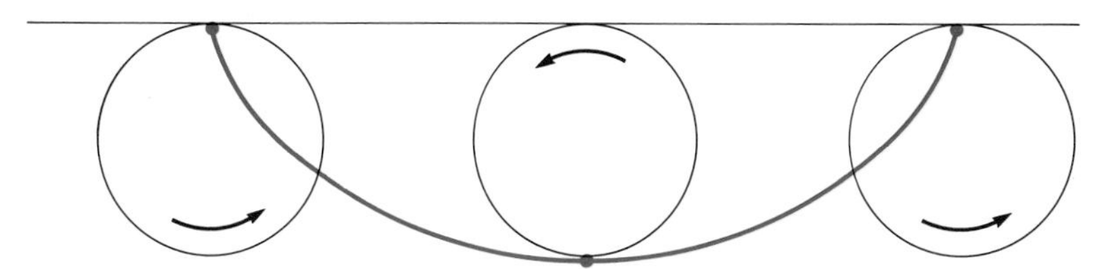

Cycloid

The cycloid is the curve traced by a point marked on the circumference of a circle as it rolls along under a horizontal straight line. This curve solves the problem of the brachistochrone, *posed in 1697 by Jean Bernoulli (1667–1784): Connect two points at different heights in the same vertical plane by a smooth wire along which a bead can slide. Measure the time that it takes the bead to descend from the higher point to the lower one for various shapes of the wire. Which shape produces the* shortest time?

moves along a part of a *circle.* Huygens found that the swing of the pendulum could be of any size, if the bob moved along a *cycloid,* which is the path followed by a point marked on a circle as it rolls along under a horizontal line.

Uncontrollable, external events make us aware of the passage of time: The Sun rises and sets; the stars revolve in the heavens; the seasons come and go. Consider the day, as an example of a measure of time. From noon, when the Sun is at its highest point, to the next noon is a *solar* day. Instead of choosing the Sun, one can pick a star, observing it at its highest point in the sky on consecutive nights. That interval is a *sidereal* day.* The two days are not quite equal, however: the solar day is about four minutes longer.

Why should there be a difference? After all, the Sun is also a star. Yes, but the Earth happens to be in orbit about *that* star. To see that this matters, it helps to look at an extreme example: A hypothetical planet "Earth" revolves about the "Sun" with a period of one year. It also rotates, in the same sense as its revolution, with a *sidereal* day of one year. How long is its *solar* day?

We begin at noon in "Los Angeles" on January 1. Now, if the "Earth" were *not* rotating relative to the stars (this is situation A in the upper diagram on page 43), then a quarter of a year later, on April 1, the "Sun" would be setting in the east in "Los Angeles." But "Earth" *is* rotating relative to the stars, taking a year for one rotation (situation B). The quarter of a full rotation as of April 1 has

*This definition is slightly different from that of the astronomers, who like to use the rotation axis of the Earth (the line between the poles) as a reference line. The rotation axis does not hold steady in space, however, a fact already known (if not in these terms) to the Greek astronomer Hipparchus in the second century B.C. The spinning of the Earth produces a slight bulge at the equator, which gives the gravitational fields of the Moon and the Sun something to work on, particularly because the rotation axis is not perpendicular to the plane of the Earth's orbital motion. This causes the axis to change direction, to *precess,* like a spinning top displaced from the vertical position. In about 10,000 years, that precession will bring the rotation axis more in line with the bright star Vega than with the Cynosure, the present pole star, which is pointed out by the Big Dipper (see the star map on page 42).

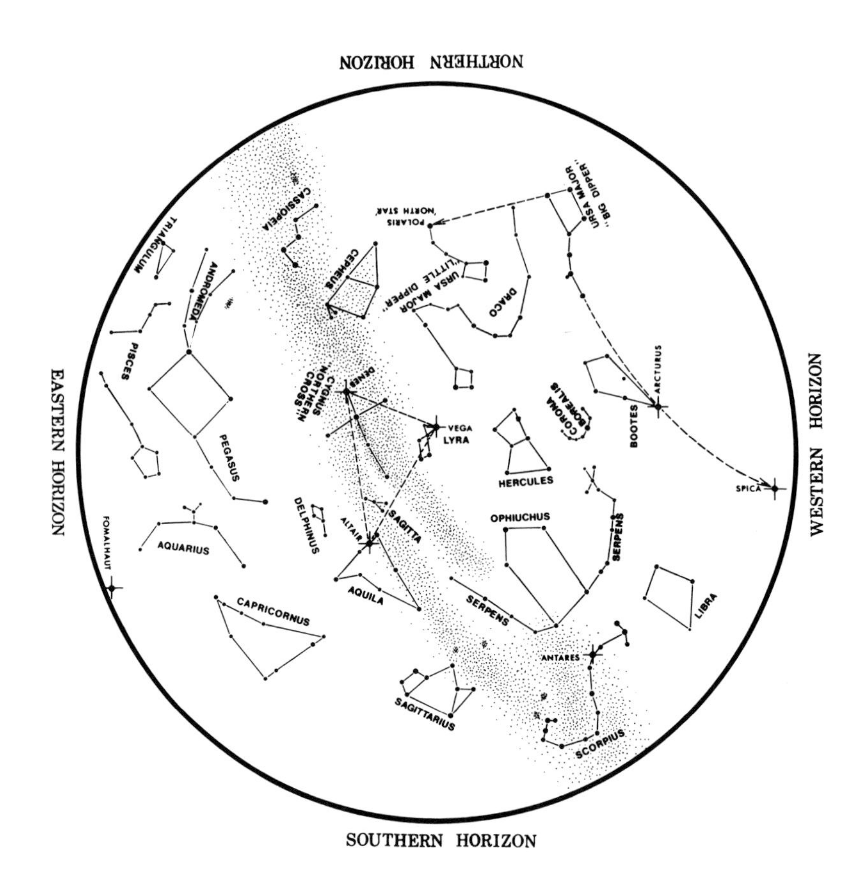

The night sky in August

Look for Vega near the center of this star chart; Polaris is above, and the Big Dipper is at its right.

kept "Los Angeles" in full sunlight—it is still noon. Indeed, it is always noon in "Los Angeles"; the solar day is forever. Incidentally, this hypothetical relation between the "Sun" and the "Earth" is just that between the Earth and its orbiting Moon, which eternally presents the same face to Earth.

This example illustrates the fact that in one year the number of rotations of a satellite, relative to the body about which it is revolving in the same sense, is one less than the number of rotations relative to the distant stars. The year of $365\frac{1}{4}$ solar days has $366\frac{1}{4}$ sidereal days—that is, a solar day is longer than the sidereal day by about one part in 365. Now a twenty-four-hour day has $24 \times 60 = 4 \times 360$ minutes. Dividing this by 365 tells us that a solar day is almost four minutes longer than the sidereal day. To be precise, this is the *mean solar day*.

Solar day for "Earth" (right)

In situation A, the "Earth" is not rotating relative to the stars. In situation B, the "Earth" is rotating relative to the stars in the same sense as its revolution about the "Sun," taking a year for one rotation.

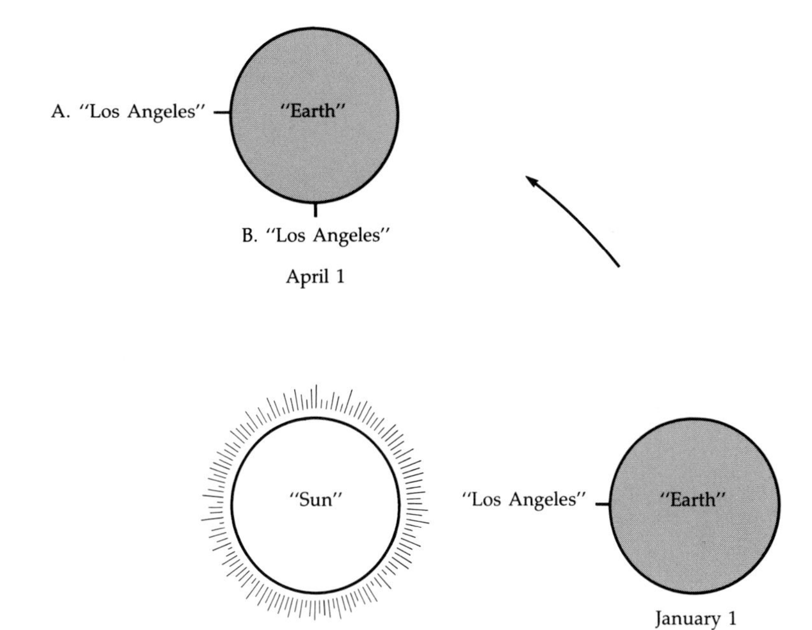

Solar day for Earth (below)

The Earth revolves full circle (360°) about the Sun in slightly more than 360 days. It therefore moves through an angle of about 1° in a day. Let us start at noon in Los Angeles and sight (indirectly) on a star in the same direction as the Sun. After one sidereal day, that star will again be at the same position in the sky. But not so the Sun, which will be slightly displaced to the east. The Earth must rotate through one additional degree before it is exactly noon on that day. Because it takes 24 hours = 360 × 4 minutes to rotate 360°, that extra degree needs four minutes. The solar day is about four minutes longer than the sidereal day.

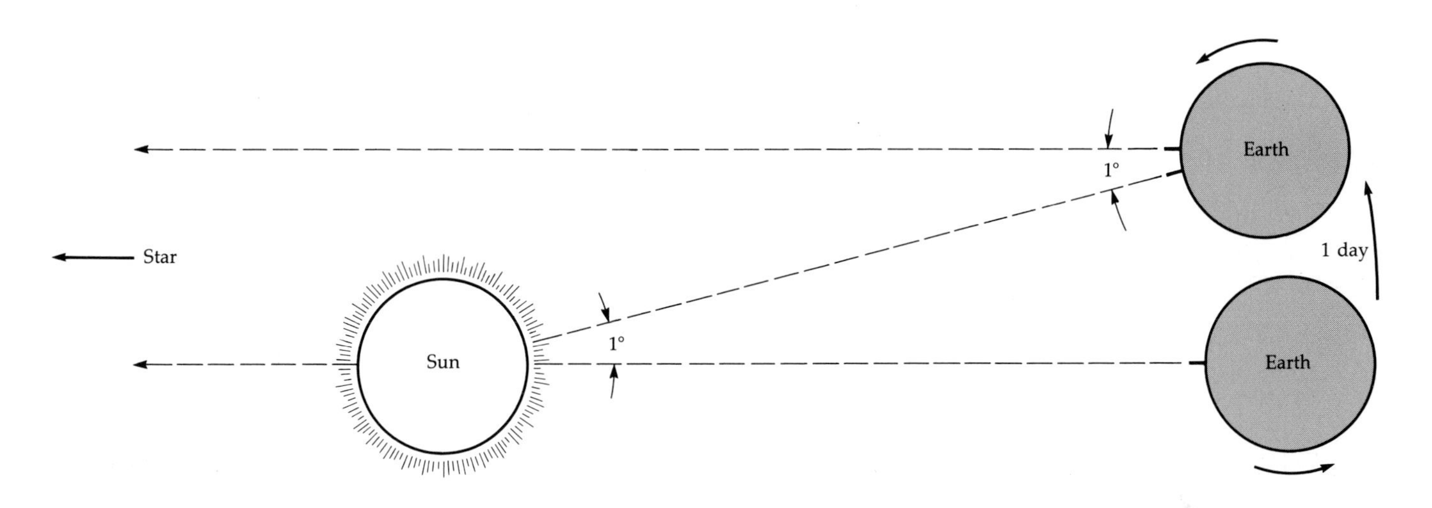

The problem with using the solar day as a standard of time is not that it is unequal to the sidereal day but that *the* solar day does not really exist. Its duration varies from day to day. There are two effects at work here. First, the Earth does not move in a circle about the Sun but, as Kepler discovered, in an ellipse, with a varying distance and speed relative to the Sun. Second, the Earth's rotation axis is not perpendicular to the plane of the Earth's orbital motion about the Sun. In consequence, the orbital motion and the rotational motion add up in different ways at various parts of the orbit. The combination of both effects makes the solar day vary in length by as much as a minute in the course of a year. It is not unreasonable, then, to define the mean solar day as the average of the solar day over the year, which is what our calculation gives.

How constant in time is the sidereal day? Theory has something to say about that. It is a consequence of Newton's (and Einstein's) mechanical equations of motion that a rigid spinning body isolated in space, one without any forces acting on it, *rotates* at a constant rate relative to an inertial reference frame, the frame of reference used by an inertial observer. As mentioned in Chapter 1, that is a reference frame in which isolated bodies *move* in straight lines. Experience teaches that inertial reference frames are those in which the background of distant stars is not rotating; so inertial frames are the reference frames of the sidereal day. Thus, a rigid, isolated Earth will maintain its rotational speed—its sidereal day—indefinitely.

But Earth is not isolated; it orbits the Sun, and has the Moon for company. Nor is it rigid. The oceans are affected by the gravita-

Bay of Fundy, high and low tides

Mount St. Helens erupting (above left), and earthquake damage in Los Angeles (above right)

tional pulls of the Moon and the Sun, which raise tides that dissipate energy through friction. The instability of the Earth's crust is quite evident in volcanic activity and earthquakes, and there are seasonal movements of vast amounts of snow and ice. As a consequence, although the rotating Earth is a superb keeper of sidereal time, it should not be a perfect one. The first evidence for that came from the records of ancient solar eclipses.

The time and place of such events, in which Earth, Moon, and Sun are lined up, can be computed with great precision from Newtonian gravitational theory. In particular, the eclipses of antiquity can be retroactively predicted and compared with the records, under the assumption of a constant sidereal day. There is a discrepancy, however. As compared with this prediction, the actual location of an eclipse that occurred, say, twenty centuries ago, is displaced to the east by about 45° of longitude (one-eighth of the circumference of the Earth at a given latitude). Therefore, the Earth, which rotates from west to east, must have been spinning more rapidly in earlier times, which means that the sidereal day was shorter then. The lengthening of the sidereal day with the passage of time works out to be $\frac{1}{600}$ of a second each century.

The need for a better time marker than the sidereal day has been met, in recent years, by atomic clocks (see Chapter 4). These extremely accurate devices have unveiled such tiny changes in the Earth's spin as those produced by the shifting of ice and snow. To

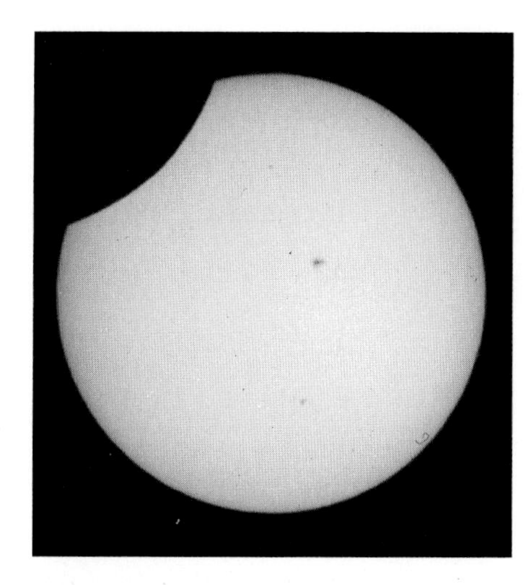

The beginning of a solar eclipse

keep the errors in marking time that are a result of such changes from accumulating, leap seconds are now introduced, usually one each year.

BOX 2·3 ***The Leap Second***

International Herald Tribune, *June 20–21, 1981.*

June 30 Will Have An Extra Second

Reuters

WASHINGTON — June 30 will be one second longer this year to get in step with the earth's rotation, the U.S. Commerce department said.

It explained that the earth's rotation, on which solar time is based, is not as regular as man-made atomic timepieces. These clocks are used by scientists in experiments and are more constant than the earth's spin.

Since 1972 scientists have introduced so-called leap seconds to keep atomic clocks from getting ahead or behind solar time. The department said that in most years, a single leap second, added in December, is a enough for adjustment purposes. Instead of waiting until December this year, the International Time Bureau based in Paris decided to make the adjustment on June 30, just before midnight.

RELATIVISTIC TIME

In the preceding section, time was treated as though it were the same everywhere. Granted, the time in New York differs by three hours from the time in Los Angeles, but that is a practical difference, not a fundamental one. It has been so designed that, within a succession of one-hour time zones, local time everywhere keeps about the same relation to the Sun's position in the sky. When clocks are set with the aid of a radio signal, they will be set very nearly the same, even for listeners separated by many hundreds of kilometers.

The situation changes, however, when we deal with astronomical distances. An electromagnetic signal (radio or light) sent to the Moon and reflected back to Earth arrives about two and one-half seconds later. Such time delays would make conversation with an astronaut on Mars difficult, indeed: even if Mars were at its nearest point to the Earth, about seven minutes would elapse before receiving the answer to a question.* In the televised account of the November 1980 flyby of Saturn, there came the dramatic announcement that the point of closest approach to the giant planet had been reached. But the signal carrying the picture that Voyager had taken at that time did not arrive on Earth for about an hour and a half.

What does it mean, then, to set a clock by another clock (to synchronize them) when the two clocks are far apart in space? On Earth, you look at a clock on the wall and set your own clock to the same time. But suppose that clock on the wall is within a spacecraft sitting on Mars and you are looking at a picture of it on a television screen. What do you do then?

To answer that, let's consider the example of Mission Control on Earth (E) and an astronaut on Mars (M), when the round-trip time for electromagnetic signals (light) happens to be ten minutes. Both E and M know this because each location is fitted with a device that reflects electromagnetic signals back along their path, and each observer, at his own location, has just measured the round-trip time. There is also a television link that enables the two observers to view the other's clock.

Suppose that M sees E's clock on the screen and it reads 12:00. Does he set his own clock at 12:00? No, he knows that the light

---

*Telephone conversations on Earth itself are troublesome when they are relayed by geosynchronous satellites. Stationed in orbit over the equator at an altitude of 36,000 km, such satellites revolve about the Earth in exactly one day, thus holding steady over any chosen point on the rotating Earth. A relayed question and response travels at least 144,000 km, which produces a time delay of one-half second or more. That is enough to throw off the normal pattern of speech.

waves took five minutes to reach him; so he sets his clock at 12:05. Back on Earth E sees that M's clock has just been set at 12:05. He looks at his own clock and it reads 12:10. He nods approval. The two clocks have been synchronized. Each observer, allowing for the travel time of the connecting signal, agrees that the two clocks read the same time.

All this can be accomplished within the limits of present-day space-age technology and experience. But what if we imaginatively expand the technological limits a bit? When Mars and Earth are closest to each other, their relative speed is about 10 km/s (for reference, this is not much greater than the speed of a satellite in low Earth orbit). We want to be able to ignore this relative speed and regard Earth and Mars as being separated by a fixed distance, from the viewpoint of a third observer (S) in a spaceship. To that end, we give the spaceship a much larger speed of, say, 1,000 km/s, which is still small compared with the speed of light (300,000 km/s).

Now suppose that, at a certain time on E's clock, which can conveniently be called time zero, E sends a short signal, a pulse, to Mars. Then, allowing for how long it took the information to reach him on Earth, he records his own time for the arrival of the signal at Mars. To E, the arrival of the pulse is an *event,* something that happens at a certain distance away at a certain time. That distance and time (which is the time required for the movement of the light pulse from Earth to Mars) are related by the speed of light.

Before leaving Earth, S synchronized his clock with that of E. Then, at the instant that E sends the pulse, a time that both observers agreed to call time zero, the spaceship heads for Mars at the given speed. How does S describe the event that is the arrival of the pulse at Mars? To this observer, Mars is approaching; therefore, the light pulse has a shorter distance to travel. So S records a shorter distance to the event than E does. Then he must also record an earlier time, because the ratio of distance to travel time, the speed of the light pulse, is the *same* for the two observers in uniform relative motion. We conclude that two such observers do *not* assign the same time to a distant event.

Unlike the difference between Eastern and Pacific standard times, this is a *fundamental* time difference, which sets limits to the physics that Newton erected on the concept of absolute time. Yet, in choosing a speed for the spaceship that is small compared with that of light, (in the ratio 1:300), we are still close to Newtonian ideas, for the difference between the two times is a small fraction of either one (in the ratio 1:300), the spaceship having traveled only that small fraction of the distance between Earth and Mars before the pulse reached Mars.

A slight twist in this arrangement gives us some practical results. Suppose that M sends a light signal of a certain frequency toward Earth. This signal is viewed by E and by S, who is in motion relative to E and M. Observer E records the electromagnetic oscillations at a fixed point and finds the frequency of the train of waves. Then, at a given time, he measures the distance between adjacent wave crests to get the wavelength.

How does S describe the same wave train? Relative to E, he is advancing to meet the waves: they arrive at shorter intervals, which is to say, a higher frequency. If S were moving in the *opposite* direction, away from Mars, the light waves would arrive at longer intervals; he would detect a lower frequency. The same thing happens with water waves and sound waves, a phenomenon known as the *Doppler effect:* if you move rapidly past some vibrating object, the frequency of the detected waves drops abruptly as motion toward the source becomes motion away from it.

The Doppler effect displays a similarity between light and other types of waves, but there is a fundamental difference. As pointed out in Chapter 1, in the example of water waves, an observer moving toward a source of waves finds their speed to be increased; when he moves away, in the same direction as that of the waves, their speed is decreased. Indeed, he can catch up with the waves, after which they are not moving at all. Next, recall that the speed of a wave is the product of frequency and wavelength. The observer's motion away from the source lowers the speed of the waves and it lowers the frequency, which is consistent with the wavelength being *independent* of the observer's motion. If, in particular, the observer moves with the waves, the relative speed is zero, and the frequency is zero because no waves are passing by.

But this is *not* the way light behaves. One cannot overtake light; its speed is unaffected by the motion of the observer. We conclude that, when the *frequency* is *increased* in consequence of observer motion, the *wavelength* is proportionally *decreased* so that the product of the two remains equal to the speed of light.

To understand how this wavelength change comes about, let us return to S, moving toward Mars, who is about to measure the wavelength of the light emitted from Mars. He begins by observing the two events that, to E, are the positions of two adjacent crests at the same time. To S, however, these events are *not* at the same time because they are at different distances from him. (For simplicity we take the nearer crest to be at his position.) As we have learned, the farther crest will be assigned an earlier time. Therefore, to carry out his wavelength measurement, he must shift his observation of the farther crest to the *later* time at which he observes the nearer crest. During that additional time interval the farther crest *moves* closer to S. And so, indeed, S, traveling toward Mars, will measure a shorter wavelength than E does. (In reality, S does not go through the two stages described here to compare his findings with those of E; he simply measures the distance between adjacent crests at a given time.) Inasmuch as the wavelength change involves the *motion* of the waves, it will be much more significant for ultrafast light than for slowly moving water and sound waves, just as we know it to be.

Incidentally, the Austrian physicist Christian Doppler (1803–1853) proposed in 1842 that the frequency effect be applied to light, but it was not until the development of photographic spectroscopy by Sir William Huggins (1824–1910) in 1868 that its reality could be demonstrated and used to measure the relative speeds, along the line of sight, of some stars. Not surprisingly, the Doppler effect, along with the aberration of light, was one of the topics discussed by Einstein in the epochal 1905 paper.

TIME DILATION

More will be said in Chapter 3 about how time and space have become intertwined in the theory of relativity. It is more important now to address the following question about *time:* Does a relatively moving clock beat the same period as an otherwise identical stationary one? The preceding discussion, which was restricted to relative speeds that are small compared with that of light, took for granted an affirmative answer. Let us see what happens when we remove that restriction.

In order not to bog down in the practical details of clock making, we devise an ideal clock that uses light. Think of two precisely parallel mirrors with perfect reflecting surfaces and a pulse of light that moves back and forth between the mirrors in a path at right angles to their surfaces. The time that it takes for light to pass from

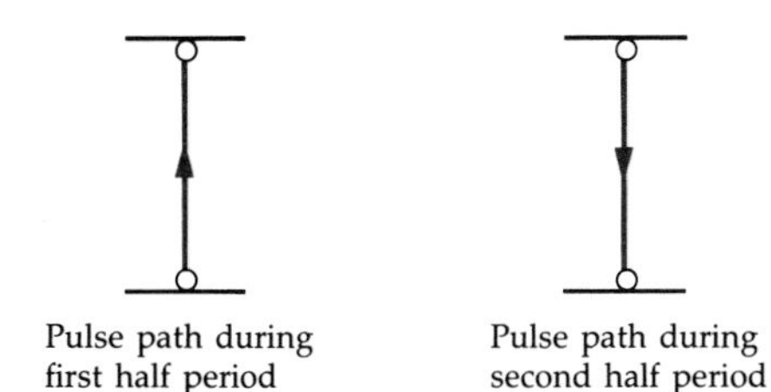

Stationary light clock

one mirror to the other and back is its *period;* that is, one tick of our perfect light clock (the time for a one-way trip being one-half period). Because the mirrors are perfect, the light pulse bounces back and forth, continuing its precise ticking forever. All observers can build identical clocks by comparing their mirror systems and assuring that the distance between parallel mirrors is precisely the same.

Now suppose that two observers, E and S, build such identical light clocks and carefully compare them in the laboratory, ascertaining that they tick in unison (that they are synchronized). Then let S take his apparatus aboard a spaceship of the far future, one that can move past E at a good fraction of the speed of light. Observer S arranges the surfaces of his mirrors to be parallel with the spaceship's direction of motion. As far as S is concerned, when the spaceship moves uniformly, everything behaves exactly as in the laboratory; the light pulse moves back and forth vertically (perpendicular to the mirror surfaces) and the clock ticks on as before.

But not so to experimenter E! Whereas S says that the light pulse went from a spot on the bottom mirror *straight up* to a point on the top mirror directly above that spot, E observes S's whole apparatus moving, say to the right, during the time that it takes the light to reach the upper mirror. E agrees with S about the place at which the pulse strikes the upper mirror but, because the apparatus is moving, that point is no longer where it was when the pulse left the lower mirror. (This is analogous to the airplane of Chapter 1, blown off course by a wind.) According to E, then, the light pulse traveled upward and to the right along a slanting path, and thus traveled a *path that was longer than the perpendicular distance between the mirrors.* So E and S disagree on how far the pulse went between reflections. But we know that they must agree on the speed of the pulse. Hence, they must also disagree on how long it took—that is, on the period of the clock. S's clock, according to E, has a longer period; it ticks *more slowly.*

But this is not the way S sees things; relative to him, E is moving by, to his right. Whereas E says his light pulse moves directly

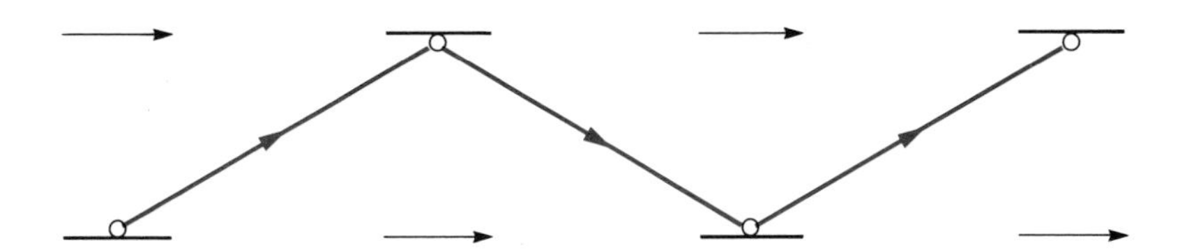

Observer E's view of S's clock, and S's view of E's clock

upward between his mirrors, S sees E's pulse moving on a longer slanting line to the upper right. Therefore S says that E's clock ticks more slowly.

If two observers are in uniform, unaccelerated motion relative to each other, neither speeding up, slowing down, nor changing direction, and they are carrying identical clocks, each will observe the other's clock to be running more slowly. Is one of them right and the other one wrong? No, both are right. Einstein's principle of relativity insists that two such observers are on exactly the same footing, that each one is reporting an objective fact that can be verified by any other observer in the same circumstances.

But couldn't a clock constructed another way (a mechanical clock, for example) behave differently? No, it could not; that is contrary to the principle of relativity. When we insist that only relative motions are physically significant, we are saying that nothing done by S within his uniformly moving spaceship (as with Galileo below decks in a uniformly moving ship) could disclose the common speed with which S and his spaceship are moving or, indeed, that there *is* any such motion. Suppose, then, that two differently made clocks are synchronized on Earth and placed aboard the spaceship, which is then boosted off Earth to move uniformly at high speed. If motion did have a different effect on the two clocks, causing them to fall out of synchronization, that would tell S that he was moving and would violate the relativity principle. In short, all clocks must slow down in the same way when observed in relative motion. This is not a property of any special device, but of *time itself.*

Very well, a relatively moving clock is observed to beat a longer period, to tick more slowly. But by how much? The light clock at rest (relative to the observer) has a period T. In half that time, a light pulse, traveling at speed c, crosses along the shortest, perpendicular line between the plates, a distance labeled A. Now the clock is observed in motion at speed v. As we have discovered, the moving clock has a longer period, called T'. In one-half of this interval a light pulse, traveling at speed c, crosses along a slant line between the plates, a distance labeled C. Both A and C are traversed at speed c. Therefore, the ratio of these distances is the same as the ratio of the times taken to travel them (moving at a fixed speed for double the time takes you double the distance) or, in symbols,

$$\frac{A}{C} = \frac{T}{T'}. \tag{2.1}$$

While the light pulse goes the distance C, the mirrors of the clock, moving at speed v, travel a distance called B. The ratio of two distances traveled in the same interval is the same as the ratio of the speeds with which the distances are covered (moving for a given time at double speed doubles the distance covered) or, in symbols,

$$\frac{B}{C} = \frac{v}{c}. \tag{2.2}$$

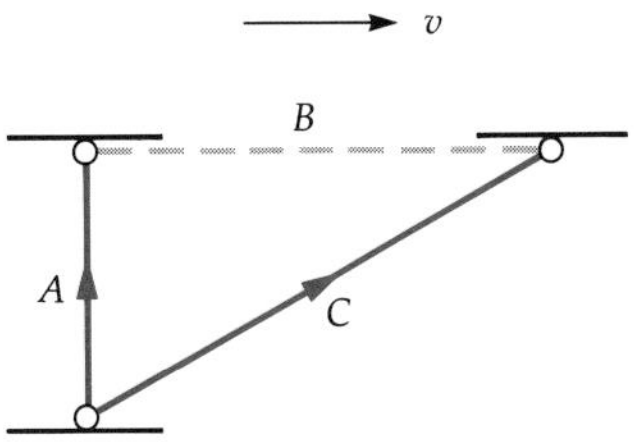

Now let us consider three points at which our light pulse touches the reflecting plates of the light clock: a starting point on the lower mirror; the point directly above it on the upper mirror of the stationary clock; and the point on the upper mirror that the pulse reaches first in the moving clock. These three points are connected by three straight lines of lengths A, B, and C, which form a right-angle triangle. As stated in Chapter 1, the three lengths are related by Pythagoras' theorem*:

$$A^2 + B^2 = C^2. \tag{2.3}$$

For our purposes, however, a more useful form of this relation is produced by dividing both sides of the equality by C^2,

$$\left(\frac{A}{C}\right)^2 + \left(\frac{B}{C}\right)^2 = 1,$$

then subtracting $(B/C)^2$ from both sides,

$$\left(\frac{A}{C}\right)^2 = 1 - \left(\frac{B}{C}\right)^2,$$

and, finally, getting A/C itself by taking the square root of each side,

$$\frac{A}{C} = \sqrt{1 - \left(\frac{B}{C}\right)^2}.$$

*Capital letters are used here to prevent confusion with the symbol for light speed.

Then if we replace A/C and B/C with the physical ratios given in relations 2.1 and 2.2, we get

$$\frac{T}{T'} = \sqrt{1 - \left(\frac{v}{c}\right)^2},$$

or, inverting both sides,

$$\frac{T'}{T} = \frac{1}{\sqrt{1 - (v/c)^2}} = \gamma. \tag{2.4}$$

This gives us the precise value of the lengthening, or dilation, of the clock's period when it is observed in motion at speed v. The Greek letter *gamma* (γ) is used as a symbol for this important measure of the effect of relative motion.

Although these explicit expressions did not appear in Chapter 1, $1/\gamma$ did appear there implicitly in connection with airplanes of

Gamma Corresponding to Various Speeds

Moving object	v	v/c	*Gamma* (γ)
Automobile	100 km/h	0.00000009	1.000000000
Concorde SST	2,000 km/h	0.000002	1.000000000
Rifle bullet	1 km/s	0.000003	1.000000000
Earth escape speed	11 km/s	0.000037	1.000000001
Orbital speed of Earth	30 km/s	0.0001	1.000000005
10% of light's speed	30,000 km/s	0.1	1.005
50% of light's speed	150,000 km/s	0.5	1.155
		0.9	2.294
		0.98	5.025
		0.988	6.474
		0.99	7.089
		0.999	22.37
		0.9992	25.00
Muons in CERN experiment		0.9994	28.87
		0.9999	70.71
		0.999999	707.1
		0.99999999	7071

airspeed *c* and a jet stream of speed *v;* it is the ratio of round-trip times, perpendicular to and parallel with the jet stream, respectively. What was said there about the ratio of round-trip times when v/c is very small can now be stated as: For very small v/c, γ differs from 1 by $\frac{1}{2}(v/c)^2$. With v equal to the speed of the Earth in its orbit, that difference was one part in 200 million, or 0.000000005. The table on the facing page presents other examples that make it clear why the relativity dilation effect plays no role in ordinary life. Speeds that are a large fraction of the speed of light must be reached before time dilation is important. Such speeds are regularly achieved in physics laboratories today but only for atomic particles. Do atomic particles have properties that can be used to mark time, thereby supplying a swiftly moving clock? All living creatures are born with a natural clock that ticks away their years of life. Atomic particles, too, can carry such a clock.

ATOMIC TIME MARKERS

Many kinds of atoms and subatomic particles are unstable; that is, they spontaneously disintegrate, or *decay,* into other particles after a certain interval. For example, the most common kind (isotope) of uranium atom (U^{238}) decays through a series of other stages, finally ending up as a lead atom. This is an example of natural *radioactivity.* The U^{238} atom happens to decay very slowly; chances are fifty-fifty that it will last for about 4.5 thousand million years; that is, after 4.5 thousand million years, half of the atoms in a sample of uranium will have decayed into something else. That period, 4.5 thousand million years, is called the *half-life* of uranium-238. After another half-life (9 thousand million years in all), half of the remaining atoms will have decayed, leaving only one-fourth of the original sample. After a third half-life, only one-eighth will be left, and so forth. By noting what fraction of the uranium atoms in an ore sample have decayed into lead, we can tell the age of the ore. In this way, by supplying a definite unit of time, radioactive atoms or particles provide a sort of atomic clock.

Another kind of unstable particle is the mu-meson, or muon. Muons were discovered by the physicists Carl Anderson and Seth Neddermeyer, and others, in the years 1936–1938. These positively or negatively charged particles turned up in the study of cosmic rays, which are the nuclei of atoms (mostly of hydrogen) that travel

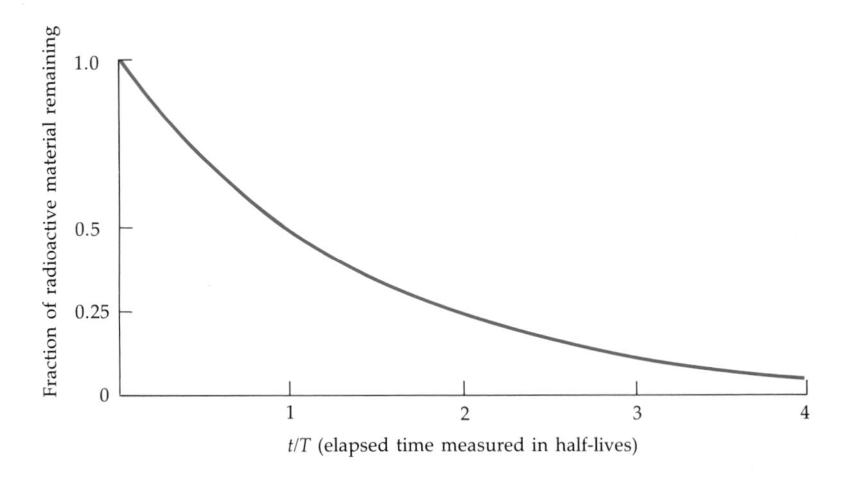

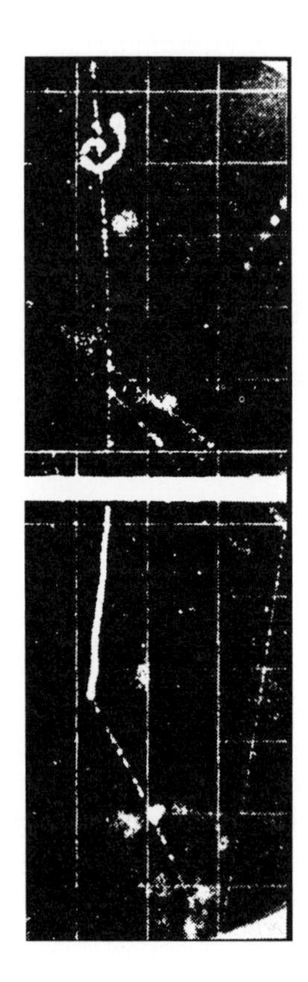

Disintegration of a meson

In this cloud-chamber photograph, a positive meson, moving downward, has penetrated an aluminum plate, thereby losing most of its energy (it becomes more heavily ionizing), and has disintegrated with the emission of a positron.

through interstellar space at nearly the speed of light. Cosmic rays continually strike the Earth, bombarding its upper atmosphere with a rate of energy influx about equal to what the Earth receives from starlight. The cosmic rays do not penetrate our atmosphere; rather they smash into air molecules and break them up into many other particles that continually rain down to the ground. The secondary particles that reach the ground in the largest numbers are the muons.

Although 206 times as massive, muons are much like electrons. They are unstable, however, and, if they are negatively charged, decay into electrons (and other particles, called neutrinos, which are much more difficult to detect) with a half-life of only 1.5 millionths of a second (1.5 microseconds, abbreviated μs). (Positively charged muons produce a particle called the positron, which will be discussed later.) Muons, therefore, provide an atomic clock with a time unit that is quite short.

Today, we can produce muons to order in the laboratory by bombarding other atoms with high-speed protons from an accelerator. Because muons are electrically charged, they are deflected in a magnetic field; so they can be stored in a donut-shaped vacuum tube surrounded by powerful electromagnets and forced to keep circling through the tube, until they decay.

In 1976, a classic experiment with muons was performed at the laboratory of the European Council for Nuclear Research (CERN) in Geneva. Muons were injected into a storage ring at the very high speed of 0.9994 times that of light. Counters stationed around the

CERN muon storage ring

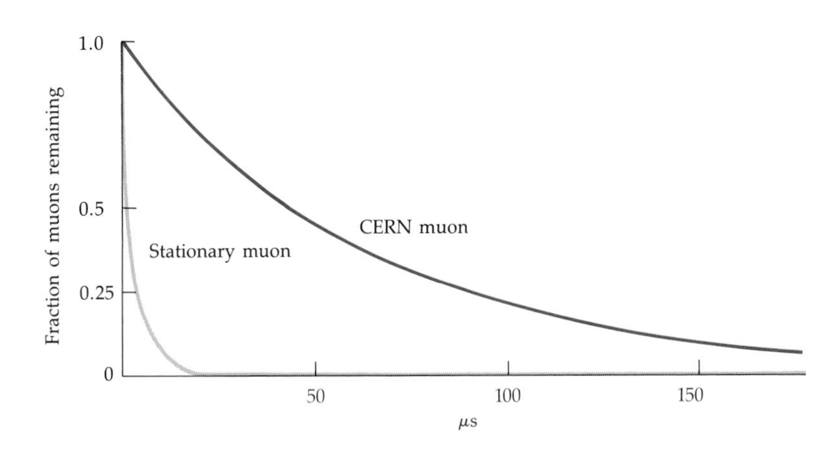

Decay curves for CERN muons and stationary muons

ring could detect the electrons produced by the muon decays and thus measure the rate of decay; that is, the half-life of the rapidly moving muons. The half-life of 1.5 μs given earlier is for stationary muons. The half-life of those in the CERN experiment was measured as 44 μs, almost thirty times as long. The table on page 54 shows that γ, for the speed of 0.9994 c, is about 29. When the precise experimental numbers were used, it turned out that the CERN experiment confirmed the prediction of relativity to within two parts in a thousand.

Before this precision was attained, the reality of the time-dilation effect had been made clear by the presence of muons *at sea level.* The primary cosmic rays create muons near the top of the atmosphere, at an altitude of some 15 km. If those muons had a half-life of 1.5 μs, despite traveling close to the speed of light (300,000 km/s or 0.3 km/μs), half of them would be gone in 0.5 km and only a thousand-millionth of them would reach the ground. This is far too few. The only possible answer to their presence in larger numbers is that fast muons age more slowly (live longer before decaying) than do muons at rest.

The muon, moving rapidly from the top of the atmosphere to the ground, lives longer by the value of the factor γ for its particular speed. (For simplicity, the fact that the electrically charged muon transfers energy to the air molecules and gradually slows down—a kind of friction effect—is ignored here.) That value might, for example, be 25, corresponding to $v/c = 0.9992$, (see the table), which gives the fast muon a half-life of $25 \times 1.5 = 37.5$ μs. Suppose that, in a thought experiment, we are moving *with* such a muon. This muon has a half-life of 1.5 μs, because it is at rest with respect to us. It is the surface of the *Earth* that is approaching at high speed. But the muon has an even chance of dying before the Earth has closed in by only 0.5 km, which makes it wildly unlikely that the muon will live long enough for the Earth to move 15 km and reach that muon. Yet the fact is that the same muon has an excellent chance of reaching the observer on the ground. What happened to the principle of relativity? The resolution of this paradox is to be found in the properties of *space.*

RELATIVISTIC SPACE

A clock in *rapid motion* relative to an observer beats a *different time* unit. The 15-km stretch from the top to the bottom of the atmosphere, as judged by an observer on the ground, is in *rapid motion* relative to the observer riding with the muon. Perhaps, then, he judges it to be a *different distance,* a smaller distance? Let's find out how much smaller it must be.

The fraction of the muons that decay during the flight from the top to the bottom of the atmosphere is an objective fact, upon which different observers must agree. That fraction is determined by the number of half-lives that have elapsed in the given travel time. We compare the two versions of this number of half-lives, one supplied by the observer on the ground, the other by the observer riding with the muon. For the ground observer, looking up a distance l to the top of the atmosphere, the time required for a muon traveling at speed v is l/v. We divide this time by the half-life of the moving muon, γT (T is the half-life of a stationary muon), to get the number $(l/\gamma) \times (1/vT)$. To the observer traveling with the muon, Earth approaches at speed v and takes the time l'/v to move the distance l'. The ratio of this elapsed time to the half-life of the stationary muon is $l' \times (1/vT)$. If the number of half-lives is to be the same for the two observers, there must be a *contraction* of length along the direction of motion given by

$$l' = l/\gamma = \sqrt{1 - \left(\frac{v}{c}\right)^2}\, l. \tag{2.5}$$

Now that we have spelled it out, let's say it more succinctly. What counts here is the ratio of travel time to muon lifetime. Because the muon-based observer uses a shorter lifetime (shorter by the factor $1/\gamma$), he must reckon the travel time to be shorter by the same factor. That is equally true of the distance traveled, inasmuch as the relevant speed is v for both observers. Therefore, a length observed in motion is shortened by the factor $1/\gamma$.

This can be illustrated using the value of γ chosen earlier ($\gamma = 25$), for which v is virtually the speed of light (0.3 km/μs). Now we say that, as observed from the muon's position, Earth moves a distance of $15/25 = 0.6$ km, which takes $0.6/0.3 = 2$ μs, or $2/1.5 = 4/3$ half-life of the *stationary* muon. Back on Earth, the moving muon needed $15/0.3 = 50$ μs to reach the ground, or $50/37.5 = 4/3$ half-life of the *fast* muon. Relativity is safe and sound.

Had we carried matters one step farther, the contraction of length in motion could also have been recognized by using the light clock. When S installed a light clock in his spaceship, he aligned the mirrors parallel to the direction of motion. Now suppose that he takes along another clock, synchronized with the first one, which he installs with the mirrors *perpendicular* to the direction of motion. As has already been emphasized, the principle of relativity requires the clocks to remain synchronized after the spaceship leaves Earth and proceeds to move uniformly at high speed. How does this work out?

As observed from Earth, the light pulse in the second clock moves back and forth between the two moving mirrors, along the line of motion. Consider the light pulse as it leaves one mirror and heads toward the other mirror, which is advancing toward the pulse. That trip is shorter than it is when the clock is stationary. Then, after reflection, the pulse chases a retreating mirror; that trip takes longer. All this is analogous to the travel times of an airplane moving with, and then against, the wind. We know that the round-trip time perpendicular to the wind (call it T') is shorter than the round-trip time parallel to the wind (T'') by the factor $1/\gamma$. Equivalently, T'' is longer than T' by the factor γ. If that were the whole story, the ticking of a moving light clock would depend on its spatial orientation relative to the direction of motion, which is a violation of the principle of relativity. That violation is prevented by the *contraction* of length along the direction of motion, which here, is the distance between the mirrors. The shortening of that distance by the factor $1/\gamma$ reduces the travel time T'' by the same factor. Now, $T'' = T'$, as it should.*

Emphasis should be placed on the phrase just used for length contraction: "along the direction of motion." When we first considered the light clock in motion parallel to the reflecting plates, it did not occur to us to question the value of a length *perpendicular* to the direction of motion—namely, the distance between the mirrors. That implicit assumption is correct; the harmony with the relativity principle that has just been achieved would be upset were we to suppose that relative motion affects lengths perpendicular to that motion.†

---

*This is a relativistic recapitulation of a pre-Einsteinian episode. In 1889, the Irish physicist George Fitzgerald, seeking to explain the failure of the Michelson-Morley experiment to detect a motion of the Earth relative to the ether, suggested that the lengths of all moving bodies are contracted in the direction of motion by the factor $1/\gamma$. It was recognized later that the failure of an analogous experiment, in which the paths were of different length, implied as well the dilation of time. Einstein's theory of relativity removes the need for such special hypotheses.

†The reader is invited to assume that such a length *is* changed, by a numerical factor depending on relative speed, and to repeat the chain of arguments using the light clock and the muon. Now is $T'' = T'$?

RELATIVISTIC SPACE-TIME

As we have just seen, the relativity of time—the fact that any temporal measure is relative to the observer—implied an analogous relativity of spatial measures. The absolute time and space of Newtonian physics crumbles completely at speeds near that of light. Perhaps this makes you uneasy because you think that two inertial observers, reporting different impressions of the *same reality*, should agree about something. But we already know that they do: they agree on the speed of light and on lengths perpendicular to the direction of relative motion. What *is* missing, however, is a single fundamental relation from which we can infer both these *absolute* properties and the behavior of measures that are *relative* to the observer. The clue that we need to find that relation comes from what we know about moving clocks: $T'/T = \gamma$, or $T/T' = 1/\gamma$.

The square of the last relation is the equality

$$\left(\frac{T}{T'}\right)^2 = 1 - \left(\frac{v}{c}\right)^2.$$

Now let us multiply both sides by $(cT')^2$ to get

$$(cT)^2 = (cT')^2 - (vT')^2.$$

(This is just a return to the Pythagorean relation used in the light-clock discussion.) We recognize in vT' the distance L' that the *moving* clock travels in time T'. Of course, the *stationary* clock doesn't go anywhere in time T; that distance is $L = 0$. And so, by introducing $L^2 = 0$ and $L'^2 = (vT')^2$ into the last relation, we can write it as

$$(cT)^2 - L^2 = (cT')^2 - L'^2. \qquad (2.6)$$

Here, having the same value for both observers, is a quantity that is neither a space measure alone nor a time measure alone: it is a *space-time measure*.

We are going to accept this as a *general* statement about any two inertial observers, O and O′, in uniform relative motion. Each observer, employing a clock and a spatial reference frame, assigns to any event a time and a distance. (It is understood that O and O′ agree on the units used to express spatial and temporal intervals. For convenience, an event labeled $T = 0$, $L = 0$ by O is labeled $T' = 0$, $L' = 0$ by O′.) In general, cT is not equal to cT', and L is not equal to L': these are relative measures. But the two observers do agree on the value of the *difference* of their squares, as given in relation 2.6: this is something absolute. Now we shall see that this single absolute property contains everything we have learned (and more).*

Speed of Light

First, this property goes to the heart of the absoluteness of the speed of light. If observer O finds a certain motion to be at speed c, as described by $L = cT$ (the distance traveled equals speed c multiplied by the elapsed time), he is asserting that the left-hand side of relation 2.6 is zero. Then, the necessarily zero value of the right-hand side of that relation says that observer O′ will discover that the (positive) distance $L' = cT'$; that is, he also finds the motion to be at the speed c.

Time

Next, consider time dilation. (This will reverse the argument leading to relation 2.4.) Observer O's clock is located at $L = 0$. To observer O′, in relative motion at speed v, the position of O's clock is given by $L' = vT'$. On putting this information into relation 2.6, we are told that

$$\begin{aligned}(cT)^2 &= (cT')^2 - (vT')^2\\ &= (cT')^2 - \left(\frac{v}{c}cT'\right)^2\\ &= \left[1 - \left(\frac{v}{c}\right)^2\right](cT')^2,\end{aligned}$$

from which it follows that $T' = \gamma T$.

Space

To deal with the relativistic aspects of spatial measures, we must first recognize that the line along which O and O′ are in relative motion is a special direction in space. Any displacement of length L, for example, can be analyzed into its projection parallel to that special direction (we call the length of this line $L_{\parallel}$) and the projection perpendicular to that direction (its length is called $L_{\perp}$). The

*The reader who is willing to grant this without further ado is invited to "play through" to the next section. However, relation 2.11 and its context are used in Chapter 3.

three lines (of lengths L, $L_{\|}$, and $L_{\perp}$) form a right-angle triangle. Pythagoras' theorem therefore tells us that

$$L^2 = L_{\|}^2 + L_{\perp}^2,$$

and similarly,

$$L'^2 = L'_{\|}{}^2 + L'_{\perp}{}^2.$$

We use this information to rewrite relation 2.6 as

$$(cT)^2 - L_{\|}^2 - L_{\perp}^2 = (cT')^2 - L_{\|}'^2 - L_{\perp}'^2.$$

Time measures motion, which, here, is the relative motion of the two observers. And the effect of that motion is a change in time of distances *along* the direction of motion. This tells us that the above equality can be split into two independent parts, one referring to time and distances along the direction of motion:

$$(cT)^2 - L_{\|}^2 = (cT')^2 - L_{\|}'^2, \tag{2.7}$$

the other referring to distances perpendicular to the direction of motion:

$$L_{\perp}^2 = L_{\perp}'^2.$$

This last relation says something that we already know: both observers agree on lengths perpendicular to the direction of relative motion.

To help in dealing with time and lengths along the direction of motion, note that the difference of the squares of any two numbers equals the product of their sum and their difference. For example,

$$(5)^2 - (4)^2 = (5 + 4) \times (5 - 4),$$

which is indeed equal to $9 = (3)^2$, as required by the Pythagorean relation. If these numbers are replaced by symbols, the equality reads:

$$a^2 - b^2 = (a + b) \times (a - b), \tag{2.8}$$

which, for example, enables us to write

$$1 - \left(\frac{v}{c}\right)^2 = \left(1 + \frac{v}{c}\right) \times \left(1 - \frac{v}{c}\right).$$

We now apply relation 2.8 to each side of relation 2.7 and get (omitting the multiplication signs):

$$(cT + L_{\|})\,(cT - L_{\|}) = (cT' + L_{\|}')\,(cT' - L_{\|}'). \tag{2.9}$$

Notice that, when we write $L_{\|}^2 = L_{\|} \times L_{\|}$, either a positive or a negative value can be used for $L_{\|}$. We are therefore free to agree that a positive value is a displacement to the right along the line of relative motion, whereas a negative value is a displacement to the left.

Let's begin the discussion of relation 2.9 by considering a situation in which the two observers are *not* in relative motion; in effect, they are the *same* observer. Then we have $cT = cT'$ and $L_{\|} = L_{\|}'$, or $cT + L_{\|} = cT' + L_{\|}'$ and $cT - L_{\|} = cT' - L_{\|}'$, which certainly satisfies relation 2.9. But that relation continues to be satisfied if we introduce compensating numerical factors, according to

$$cT + L_{\|} = V'\,(cT' + L_{\|}') \tag{2.10}$$
$$cT - L_{\|} = \frac{1}{V'}\,(cT' - L_{\|}').$$

Indeed, when we multiply the right-hand sides, the additional factors combine into $V' \times (1/V') = 1$. We shall find that the ratio of these two statements,

$$\frac{cT + L_{\|}}{cT - L_{\|}} = V'^2\,\frac{cT' + L_{\|}'}{cT' - L_{\|}'}, \tag{2.11}$$

is quite useful.

To learn the meaning of V', let O′ move relative to O with velocity v (for positive v, O′'s relative motion is to the right; it is to the left for negative v). From O's viewpoint, then, O′'s motion is described by $L_{\|} = vT = (v/c)cT$, whereas O′ specifies his own position by $L_{\|}' = O$. When we put this information into relation 2.11 (and cancel, between numerator and denominator, cT, on the left side, and cT' on the right side), we get

$$\frac{1 + (v/c)}{1 - (v/c)} = V'^2, \tag{2.12}$$

or

$$V' = \sqrt{\frac{1 + (v/c)}{1 - (v/c)}}. \tag{2.13}$$

So V' is a measure of relative motion that is related to, but is distinct from, the relative velocity (more about this shortly). In the absence of relative motion (v = O), relation 2.13 must produce the simple result: $V' = 1$, which it does.

The relations in 2.10 give O's space and time measures in terms of O′'s. But the principle of relativity tells us that the two observers are on an equal footing, and so it must also be possible to give O′'s measures in terms of O's, as

$$cT' + L_{\|}' = V\,(cT + L_{\|}), \qquad (2.14)$$
$$cT' - L_{\|}' = \frac{1}{V}\,(cT - L_{\|}),$$

in which V is constructed in the same way as is relation 2.13, from O's relative velocity. Now, if O′ moves relative to O at velocity v, then O moves relative to O′ at velocity $-v$, which is to say, when O′ moves to the right relative to O, then O moves to the left relative to O′, at the same speed. Therefore, we must have

$$V = \sqrt{\frac{1 - (v/c)}{1 + (v/c)}} = \frac{1}{V'}, \qquad (2.15)$$

which is just what appears on solving relation 2.10 for O′'s measures in terms of O's. Very nice.

Length Contraction

This brings us to the contraction of length in relative motion. We need the relation produced by subtracting the two statements in relation 2.10 and dividing by 2,

$$L_{\|} = \frac{1}{2}\left(V' - \frac{1}{V'}\right)cT' + \frac{1}{2}\left(V' + \frac{1}{V'}\right)L_{\|}', \qquad (2.16)$$

along with this simplification of $\frac{1}{2}(V' + (1/V'))$:

$$\frac{1}{2}\sqrt{\frac{1 + (v/c)}{1 - (v/c)}} + \frac{1}{2}\sqrt{\frac{1 - (v/c)}{1 + (v/c)}} = \frac{1}{\sqrt{1 - (v/c)^2}} = \gamma; \qquad (2.17)$$

it is immediately checked through multiplication of both sides by $\sqrt{1 - (v/c)^2} = \sqrt{1 + (v/c)} \times \sqrt{1 - (v/c)}$.

Consider a stretch along the direction of relative motion. Observer O is at rest relative to the stretch and measures its length as l, whereas observer O′ is in relative motion and, at a given time T', measures the length of the stretch as l'. Now apply relation 2.16 twice, to the location and time of each end of the stretch. Then subtract these two relations. The difference of the two $L_{\|}$ values is length l; the difference of the two $L_{\|}'$ values is length l'; and the difference of the two T' values is zero. Therefore (remembering relation 2.17 we have

$$l = \gamma\, l'$$

or

$$l' = (1/\gamma)l,$$

which is the known contraction of length in relative motion.

We have now seen how the space and time measures that are relative to the observer flow from *one* absolute statement; the value of $(cT)^2 - L^2$ assigned to a space-time event is the *same* for all observers.

VOYAGE TO VEGA

The following space-time story is not entirely true. Not because it is set in the future—time will take care of that—but because some practical details are brushed aside so as not to obscure the main point.

The starship Argo[3] begins an epic voyage to Vega,* twenty-six light years (LY) distant, at a time when that objective is the pole star. Aboard her are forty-nine Argonauts, one less than the planned number—at the last moment, Castor, twin brother of Pollux, has been detained on Earth. After getting up to the speed $v = 0.988c$ (about 1 percent less than the speed of light), the voyage begins. At that speed, it takes just a bit more than twenty-six years to travel 26 LY (light speed is 1 LY/yr). The return voyage requires an equal length of time, and so the round-trip travel time is about fifty-two years. Everyone on Earth, including Castor, is that much older when the Argonauts return. But, to earthlings, Argo has been traveling at the speed $v = 0.988c$ for which (see the table on p.54) $\gamma = 6.5$, very nearly. Therefore every Argonaut, including Pollux, has aged only $52/6.5 = 8$ years!

How does all this look to the Argonauts? To them, the 26-LY stretch to Vega, traversed at a speed such that $\gamma = 6.5$, appears as

*As early as 1983, it was known that Vega is orbited by cold matter, which might be a planetary system at an earlier stage than ours.[4] At last it has become possible to go and look.

26/6.5 = 4 LY. Thus, the journey takes 4 years. Another 4 years finds them back in the solar system, which means that the Argonauts have aged 8 years. But Hercules, famed for his physical prowess, protests: "We've seen *Earth* tear away and return at almost the speed of light. They must have aged more slowly than we did!" Is Hercules right? When the twins Castor and Pollux meet face-to-face, are they still the same age, or, if not, which is the elder brother?

The apparently paradoxical conflict between the two viewpoints dramatizes something already recognized by Einstein in 1905. Two observers, with initially synchronized clocks, move apart and eventually rejoin, owing to the acceleration of *one* of them. Superficially, each can regard the other as being in relative motion and can predict that the other's clock will have lagged when they rejoin and compare readings. That is the "paradox." It *seems* to be based on the special relativistic concept that two different, uniformly moving observers are on the same footing and will give equally valid accounts of phenomena. The observer on the spaceship, however, is *not* in uniform, unaccelerated motion, because he must be accelerated to *reverse direction* after having reached the objective. The special theory of relativity does not apply to such an accelerated observer (Einstein would eventually remove this restriction in his general theory of relativity, but that's another story). Therefore, the two viewpoints are *not* equivalent, and the returning astronaut, Pollux, will indeed have aged less than his stay-at-home twin, Castor.*

A more detailed discussion of the different viewpoints of Earthlings and Argonauts is given in Chapter 3.

NOTES

1. **H. Boorse and L. Motz,** eds. *The World of the Atom* (Basic Books, 1966), p. 545.
2. *Albert Einstein,* Library of Living Philosophers, vol. 7 (1949), p. 47.
3. **Thomas Bulfinch** *The Age of Fable* (Thomas Y Crowell, 1970), p. 130.
4. *Science* 221 (26 August 1983):846.

---

*Our story has an unhappy footnote. When Castor eventually passed away, the still young Pollux died of grief. Their names are given to the twin planets discovered in the Vegan system. I cannot tell you whether the Argonauts found intelligent life on those planets; only Zeus knows.

3

$E = mc^2$

THE HUMAN SIDE

One day in 1909, when Albert Einstein was still working in the Swiss patent office in Bern, he found something in the mail. But let him tell the story.

One day I received in the Patent Office in Bern a large envelope out of which there came a sheet of distinguished paper. On it in picturesque type was printed something that seemed to me impersonal and of little interest. So right away it went into the official wastepaper basket. Later, I learned that it was an invitation to the Calvin festivities and was also an announcement that I was to receive an honorary doctorate from the Geneva University. [In 1909, that University celebrated the 350th anniversary of its founding by John Calvin; it bestowed more than a hundred honorary degrees. Their letter of invitation unanswered, the university authorities had a friend of Einstein persuade him to attend the ceremonies.]

So I traveled there on the appointed day and, in the evening in the restaurant of the inn where we were staying, met some Zürich professors. Each of them now told in what capacity he was there. As I remained silent I was asked that question and had to confess that I had not the slightest idea. However, the others knew all about it and let me in on the secret. The next day I was supposed to march in the academic procession. But I had with me only my straw hat and my everyday suit. My proposal that I stay away was categorically rejected . . . .

The celebration ended with the most opulent banquet that I have ever attended in all my life. So I said to a Genevan patrician who sat next to me, "Do you know what Calvin would have done if he were still here?" When he said no and asked what I thought, I said, "He would have erected a large pyre and had us all burned because of sinful gluttony." The man uttered not another word . . . .[1]

Count Rumford (1753–1814)

Clearly, Einstein's work had attracted attention by 1909. He himself, however, was unknown in the usual haunts of scientists, although he had begun to correspond with active physicists. One such exchange concerned the world-famous equality that titles this chapter. It was in February 1908 that Einstein wrote to the German physicist (and later Nobelist) Johannes Stark, complaining that Stark had ignored his priority in "the connection between inertial mass and energy." Two days later, Stark responded with protestations of admiration. Three days after that, Einstein replied as follows:

> Even if I had not already regretted before receipt of your letter that I had followed the dictates of petty impulse in giving vent to that utterance about priority, your detailed letter really showed me that my over-sensitivity was badly out of place. People who have been privileged to contribute something to the advancement of science should not let such things becloud their joy over the fruits of common endeavor.[2]

It was also in 1908 that a former Zürich professor of Einstein, Hermann Minkowski, announced to the world that a new vision was at hand and that it had first been clearly seen by Albert Einstein. (It is alleged that Minkowski had remarked earlier to the effect that he never expected *that* student to come up with anything so smart.) In an address titled "Space and Time," which was delivered to an Assembly of German Natural Scientists and Physicians, Minkowski began with these ringing phrases:

> The views of space and time which I wish to lay before you have sprung from the soil of experimental physics, and therein lies their strength. They are radical. Henceforth space by itself, and time by itself, are doomed to fade away into mere shadows, and only a kind of union of the two will preserve an independent reality.[3]

Overblown, but memorable.

c IS FOR THE SPEED OF LIGHT

Simultaneity

Einstein's theory of relativity dethrones Newtonian absolute time—a time that is the same for all observers, however far apart and however rapidly in relative motion. Some aspects of the relativity

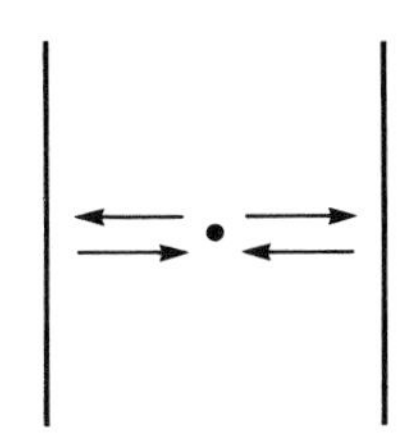

Stationary parallel plates

The source of light is centered between the plates and the arrows indicate the paths of the light.

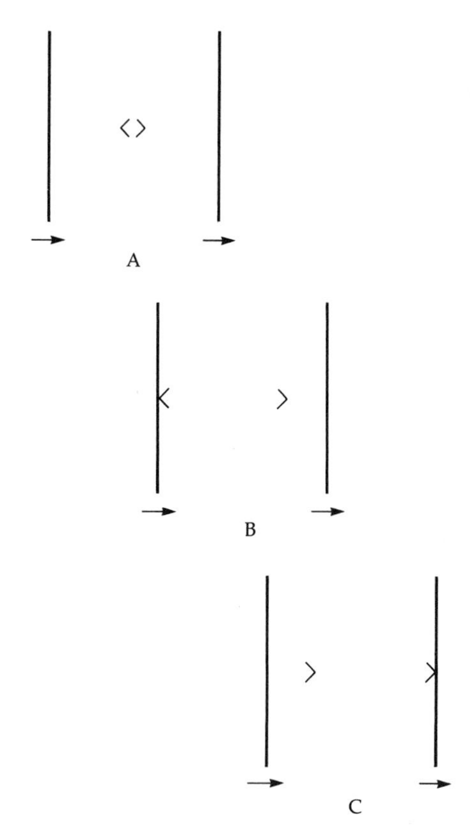

Moving parallel plates (displacement exaggerated)

A. Just after the emission of the light pulses.
B. At the first impact.
C. At the second impact.

of time were considered in Chapter 2. To develop further consequences of Einstein's non-Newtonian concepts, we return to the physical arrangement of perfectly parallel, perfectly reflecting plates that was introduced there. But now we put the source of light halfway between the plates. From it, *two* light pulses are emitted simultaneously, in opposite directions, each directed to hit the adjacent plate perpendicularly.

First, consider what is recorded by an observer at rest relative to the plates. The two light pulses are emitted at the same time, and they travel equal distances in equal amounts of time. So, he observes them to strike the two plates at the same time, *simultaneously*. After being reflected, the two pulses travel back toward the source, which they reach simultaneously. So far, so good.

Now consider a second observer who moves at a constant speed perpendicularly to the plates, parallel to the line along which the pulses move.* Relative to him the apparatus is in motion, along the line that the two pulses follow. You already know what is going to happen. The pulse moving toward an advancing plate hits it sooner than the pulse chasing a retreating plate hits that plate. The pulses do not strike the two plates at the same time—those events are *not* simultaneous for this observer. (There is nothing significant about which plate is struck first; an observer moving in the opposite direction finds that order of priority reversed.)

Now, to the next phase. The pulse that struck an advancing plate turns around to chase a retreating source; the pulse that struck a retreating plate turns around to meet an advancing source. Thus, on the return journeys everything is reversed and the two pulses reach their starting point at the same time, *simultaneously*, just as was observed when the apparatus was at rest.

The following inferences can be drawn from this example:

1. Events, at the *same* point of space, that are simultaneous for one observer are simultaneous for any other observer moving uniformly relative to the first observer.
2. Events, at *different* points of space, that are simultaneous for one observer are in general *not* simultaneous for any other observer moving uniformly relative to the first observer.

---

*Notice that, having already done this sort of thing a number of times, we dispense with a pseudorealistic context—observers in spaceships. In the simile popularized by Einstein himself, the relatively moving observer is riding in a train, one that can move at a significant fraction of the speed of light.

Time and space have become entangled. But just how far can that confusion go?

Causality

"Get your bets down. They're off! And the winner is . . . ." Here is a familiar example of events that have a definite order in time. Horses *do not* cross the finish line before they have left the starting gate. But have we not just learned that the time sequence of two events can be reversed by shifting to a new viewpoint, by changing the observer? Does that mean we can observe the finish of the race, and then, knowing the outcome, go back to bet on the winner? Has Einstein pointed the way to instant riches? Alas, no.

What connects the earlier event (the horses leaving the starting gate) with the later event (the breaking of the tape by the winner) is a physical object (a running horse). Its motion can be followed from start to finish. Any observer, at whatever relative velocity, can follow that motion and will agree on the same sequence of events. This is an example of a pair of events that have a *causal* relation, thus called because the final *effect* (the breaking of the tape) has a definite prior *cause* (the horse leaving the gate).

Things were different in our arrangement of parallel mirrors at rest that have a light source midway between them. The *simultaneous* striking of the two plates by the light pulses could not have been events in causal relation; no physical influence can travel *in-*

stantaneously from one point to the other. However, when the apparatus moved (relative to the observer), there *was* a time difference. Was there enough time to allow a physical object to move from the earlier event to the later event? No. There is an elegant way to show this by using symbols.

In Chapter 2; it was established that, whereas a time interval T between two events and the space interval L between them are relative quantities, the space-time combination

$$(cT)^2 - L^2$$

has an absolute significance; it has the same value for all (inertial) observers. Notice that such a *difference* between two positive quantities can be positive or negative, depending on which quantity is larger. It can, of course, be zero.

Suppose that we are dealing with two events in causal relation. Then the spatial interval L can be traversed, in the elapsed time T, by an object moving at a speed v that is less than, or possibly equal to, the speed of light. The distance $L = vT$ will then be less than, or possibly equal to, cT. If v is smaller than c, so that L is smaller than cT, $(cT)^2 - L^2$ is *positive.* And if v equals c, we have $L = cT$, and $(cT)^2 - L^2$ is *zero;* we have already met this description of motion at the speed of light. The absoluteness of $(cT)^2 - L^2$ means that its positive or zero value will hold for all observers of a causal relation.

We are being told, again, that a causal relation between two events is a property that is true for all observers.

Now consider two events that, to one observer, occur at the same time, but at different points of space; that is, the time interval T between these simultaneous events is $T = 0$, whereas the spatial interval between them, their distance, L, is not zero. Then $(cT)^2 - L^2$ is *negative.* The fact that it is neither positive nor zero confirms that these two events are *not* causally related. Finally, what of the second, relatively moving observer to whom these events are *not* simultaneous? No computations are needed; the absoluteness of $(cT)^2 - L^2$ assures us that this observer will find the same negative value as before. The absence of a causal relation between two events is a property that is true for all observers.

So the entanglement of space and time is very orderly indeed. If one inertial observer finds two events to be in causal relation, which implies a positive time interval T between the effect and its cause, so will all other inertial observers. This means that, as one shifts from one observer to another, the time order of cause and effect can never reverse—that is, T can never become negative. If it did, there would be some intermediate observer for whom $T = 0$, and his value of $(cT)^2 - L^2$ would be negative, which it *is not.* If, however, two events are *not* causally related and $(cT)^2 - L^2$ *is* negative, it is quite reasonable that different observers of the same scene can find different kinds of values for T, variously zero, positive, or negative, because there is no *physical* basis for a particular order of the events. Truly there is greater inner harmony here.

Relative Speed

It is the speed of light, as the highest possible physical speed, that underlies these considerations about causality. Perhaps you are still not entirely convinced about the reality of that limitation; the everyday, Newtonian concepts of motion are hard to shed. Consider, for example, an automobile moving at 60 km/h that passes another automobile going in the opposite direction at a speed of 80 km/h. Does each driver see the other car going by at the relative speed of $60 + 80 = 140$ km/h? Yes, certainly. Well, then, if two superrockets traveling at $0.6\ c$ and $0.8\ c$ pass each other going in opposite directions, is not their relative speed $0.6 + 0.8 = 1.4\ c$? No, it is not.

As we have learned, the Newtonian concepts of space and time fail for speeds approaching that of light. Inevitably, so also must the Newtonian concepts of speed, the rate at which space is traversed in time. That much is clear. What is missing for this special situation, motion along a line, are the answers to these questions:

If you do not add the individual speeds of two objects moving in opposite directions to get the relative speed, what should you do?

Does it really work out that, if one object is traveling at speed c, the relative speed of the two objects is still c?

To supply these answers, recall the discussion in Chapter 2 of the connection between the space-time measures of two observers in relative motion. Observer O′ moves relative to observer O at velocity v' (it was called v in Chapter 2, but here that symbol will have another use). As indicated in Chapter 2 (relation 2.11), their space and time measures are related by

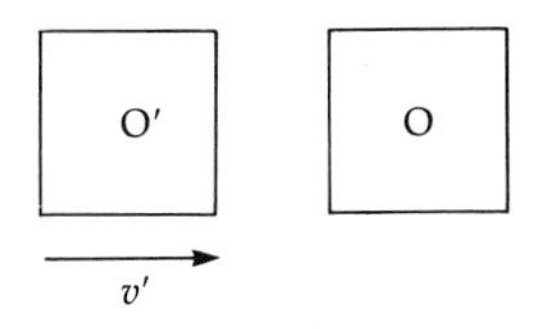

$$\frac{cT + L_{\|}}{cT - L_{\|}} = V'^2 \frac{cT' + L'_{\|}}{cT' - L'_{\|}}, \tag{3.1}$$

in which

$$V' = \sqrt{\frac{1 + (v'/c)}{1 - (v'/c)}}.$$

Now consider an object that moves at velocity v'' relative to O′. According to O′, the object moves a distance $L'_{\|}$ in time T' and $L'_{\|} = v''T' = (v''/c)cT'$. If we insert this expression for the $L'_{\|}$ in the fraction on the right side of relation 3.1 and cancel cT', we get

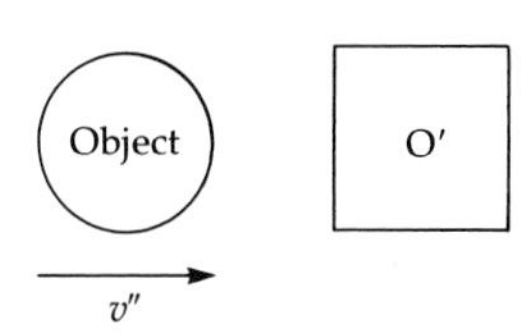

$$\frac{cT' + L'_{\|}}{cT' - L'_{\|}} = \frac{1 + (v''/c)}{1 - (v''/c)} = V''^2.$$

How fast does that object move relative to O? It moves at some velocity v through a distance $L_{\|}$ in time T, and $L_{\|} = vT = (v/c)cT$. Then,

$$\frac{cT + L_{\|}}{cT - L_{\|}} = \frac{1 + (v/c)}{1 - (v/c)} = V^2.$$

Substituting these results in relation 3.1 gives

$$V^2 = V'^2 V''^2, \tag{3.2}$$

or

$$V = V'V'',$$

which tells us how to find v.

Very pretty, but what does it have to do with the relative speed of two moving objects? Just this. As before, an object moves relative to O′ with velocity v''; a second object, observer O, has the velocity $-v'$ relative to O′ (the negative of the velocity of O′ relative to O). If v' and v'' are *positive* quantities, we have two *oppositely* moving objects. Then the desired relative speed of the two objects, which is the speed of the first object relative to observer O (the second object), is given by v. If we wish to find the relative speed of two objects moving in the *same* direction, we have only to replace $-v'$ by v'.

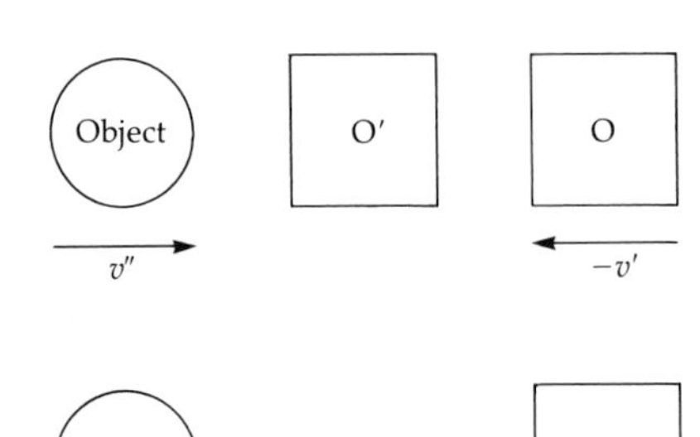

Let us take a moment to look at the relation between the two different measures of motion, v/c and V. We already know that $V = 1$ for $v/c = 0$. Now we can see, from the behavior of the denominator in V, that, as v/c increases toward the limiting value of 1, V becomes larger and larger without limit; it becomes infinitely large. For motion in the opposite direction, negative values of v/c, the behavior of the numerator in V tells us that V approaches zero as v/c tends toward -1. Thus the complete range of v/c, from -1 to 1, corresponds to V varying between zero and infinity. And, given any value of V in this interval, one can work back to the corresponding value of v/c.

To do that explicitly, we begin with

$$V^2 = \frac{1 + (v/c)}{1 - (v/c)},$$

and multiply both sides by $1 - (v/c)$ to get

$$V^2 - \frac{v}{c}V^2 = 1 + \frac{v}{c}.$$

Add $(v/c)\,V^2$ to both sides; then subtract 1 from both sides, with the result

$$V^2 - 1 = \frac{v}{c}(V^2 + 1),$$

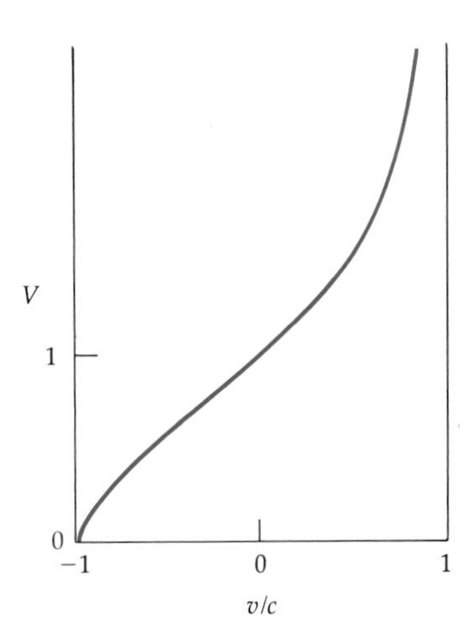

which finally produces

$$\frac{v}{c} = \frac{V^2 - 1}{V^2 + 1}. \tag{3.3}$$

You can see that, as V, and also V^2, increases from zero, goes through unity, and approaches infinity, v/c ranges from -1, through zero, toward 1, just as it should.

What happens when we consider the example of the super-rockets for which $v'/c = 0.6$ and $v''/c = 0.8$? First, we find the corresponding values of V' and V'':

$$V'^2 = \frac{1 + 0.6}{1 - 0.6} = \frac{1.6}{0.4} = 4; \text{ so } V' = 2$$

$$V''^2 = \frac{1 + 0.8}{1 - 0.8} = \frac{1.8}{0.2} = 9; \text{ so } V'' = 3.$$

Then we work out the value of V:

$$V = V'V'' = 2 \times 3 = 6.$$

Finally, we get the relative speed v from relation 3.3,

$$\frac{v}{c} = \frac{36 - 1}{36 + 1} = \frac{35}{37} = 0.946 \text{ (approximately)},$$

a relative speed that is fast but *not* faster than the speed of light.

No other outcome was possible. Given any physical values of v'/c and v''/c, numbers that lie in the interval between -1 and 1, the corresponding values of V' and V'' are numbers in the range between zero and infinity. Then V, the product of V' and V'', is also a number in that range. When this V is used in relation 3.3, the outcome is a value of v/c in the physical interval between -1 and 1. So the answer to the first question on page 75 is this:

You do not add v' and v''; you multiply V' and V''.

That brings us to the second question. Suppose that the object with velocity v'' is moving (to the right) at the speed of light: $v''/c = 1$. Is it true that v/c is also 1, no matter what v'/c may be? Yes (with the sole exception of $v'/c = -1$). The value of V'' for $v''/c = 1$ is infinitely large. The result of multiplying it by any value of V' is also infinitely large, implying that $v/c = 1$, *except* if $V' = 0$, corresponding to $v'/c = -1$. Here we meet the product of zero and infinity, which is ambiguous. (To give that product a meaning, one must specify just how the limiting values of zero and infinity are approached.) The physical situation described by $v''/c = 1$ and $v'/c = -1$ is that of two photons traveling in the *same* direction. But the concept of relative speed assumes the possibility of an observer moving along with at least one of the objects, to measure the speed of the other one. That is impossible when *both* objects move at the speed of light; relative speed becomes meaningless. It is fitting, then, that mathematical ambiguity appears just when the physical concept becomes meaningless; these are the opposite faces of a single coin.

This reference to objects moving in the same direction invites us to try a numerical example that *is* physically meaningful. An automobile traveling at 80 km/h moves away from another automobile, going 60 km/h in the same direction, at the rate of $80 - 60 = 20$ km/h. Does a rocket that zips along at $0.8c$ move away from a second rocket, traveling at $0.6c$, with the relative speed $0.8 - 0.6 = 0.2c$? Perhaps, you think that now this should be about right because it is comfortably below the speed limit of c? Let us find out.

We still have $v''/c = 0.8$ and $V'' = 3$. Now, however, v'/c is changed to -0.6. It has already been noted, in relation 2.15, that reversing the sign of v replaces V by $1/V$. Here, then, we get $V' = 1/2$ and $V = V'V'' = (1/2)3$. This gives us $V^2 = 9/4$, and

$$\frac{v}{c} = \frac{(9/4) - 1}{(9/4) + 1} = \frac{9 - 4}{9 + 4} = \frac{5}{13} = 0.385 \text{ (approximately)},$$

which is almost *twice* the Newtonian value.

This outcome is bound up with the answer to the second question: If the speed of one object is c, so also is its speed relative to any other moving object. Even for objects moving at speeds somewhat less than c, there is still a tendency for the relative speed to approach that of the faster object and be less dependent on the slower speed. Newtonian calculations of relative speed should give values, for oppositely moving objects, that are too large, and values for motion in the same direction that are too small. To make that quite explicit, we return to relation 3.2, written out as

$$\frac{1 + (v/c)}{1 - (v/c)} = \frac{1 + (v'/c)}{1 - (v'/c)} \cdot \frac{1 + (v''/c)}{1 - (v''/c)}$$

$$= \frac{[1 + (v'/c)(v''/c)] + [(v'/c) + (v''/c)]}{[1 + (v'/c)(v''/c)] - [(v'/c) + (v''/c)]} \cdot$$

After dividing the numerator and denominator by $1 + (v'/c)(v''/c)$, we can identify the relative speed:

$$v = \frac{v' + v''}{1 + (v'/c)(v''/c)} \cdot \quad (3.4)$$

Here we see plainly that, for opposing motion ($v'v''$ positive), the relative speed is less than $v' + v''$, whereas it is greater than $v' + v''$ for motion in the same direction ($v'v''$ negative). It is also clear that in the Newtonian regime, where both v'/c and v''/c are *much* less than 1—even more so their product—the relative speed is given accurately by the Newtonian sum, $v' + v''$. The reader is invited to try relation 3.4 on the numerical examples and to check that $v''/c = 1$ or $v'/c = 1$ implies $v/c = 1$.

m IS FOR MASS

All right—put in that nothing goes faster than light, and it comes back out again; there is nothing inconsistent about it. But, still, *why* cannot anything exceed the speed of light? According to Newton, a force acting on a body changes the body's momentum at a rate equal to that force. (*Momentum* is the product of *mass* and *velocity*.) If the force acts steadily, the momentum will grow at a constant rate. Eventually, the magnitude of the velocity, the speed of the body, should exceed that of light. But that does not happen; so something about the Newtonian mechanical concepts must give way as the speed of light is approached.

Conservation Laws

The challenge here is to combine Einsteinian relativistic concepts with Newtonian mechanics to produce a consistent relativistic mechanics. But, if Newtonian mechanics is not reliable, how do we know what to retain and what to discard? The answer is found in the great generalizations that grew from Newtonian mechanics but transcend it in their universality. These are the conservation laws—here, specifically, the conservation of momentum and of energy.

The conservation of momentum is implicit in Newton's three laws of motion.

1. Any body not acted on by forces retains a constant velocity and therefore a constant momentum; that is, the momentum is conserved.
2. As just mentioned, a given force produces a rate of change of momentum equal to that force.
3. The mutual forces between two bodies are equal and opposite; that is, the rate of change of one momentum is the negative of the rate of change of the other momentum, and the sum of the two momenta is constant—it is conserved.

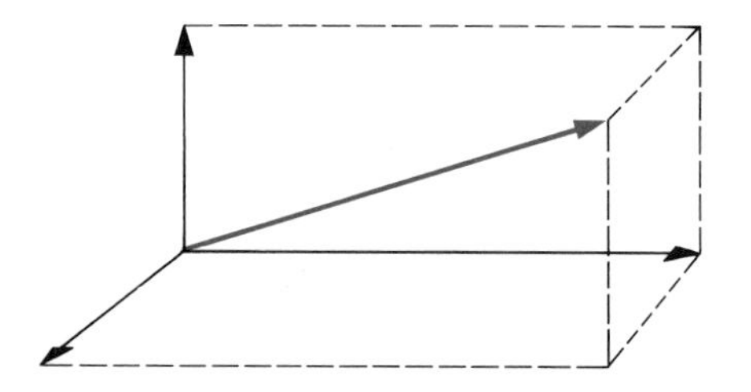

Like velocity, momentum is a directed quantity; it points along the direction of motion. The conservation laws apply separately to the *three* perpendicular directions in space into which motion can be resolved. *Energy* however, is a single, nondirected quantity. Although energy will be discussed in some detail later, it should be noted here that, for a single body, the energy of motion is a measure of the magnitude of the momentum, as determined by the speed of the body. Thus, a body acted on by no force, with a constant velocity, and speed, certainly has a constant energy.

We are going to consider elastic collisions. Think of two bodies that approach each other, make contact, and in the process become somewhat deformed, then spring back and move apart. This event is called an *elastic collision* if *all* of the initial energy of motion is transferred into the motional energy of the separating bodies; they are not left quivering or red hot, for example. An ideal example is the collision of two perfect billiard balls rolling on a perfectly smooth table. But this kind of perfection is really approached only at the atomic level. Indeed, it was the observation of elastic collisions between electrons and light, as reported in 1923 by the American physicist (and 1927 Nobelist) Arthur Compton, that proved unmistakably the particle character of light, the reality of the photon, as had been foreseen by Albert Einstein in 1905.

A Duel

Two belligerent observers, O and O′, are at rest with respect to each other, a certain distance apart. Simultaneously, they fire at each

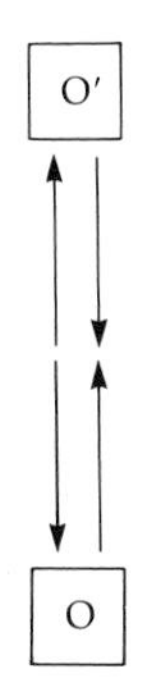

Duelists at rest, before and after collision of projectiles

other equally massive projectiles—billiard balls or atomic particles—that move at equal speeds. Thanks to unerring aim, the projectiles meet head on, halfway between the duelists,* and recoil back along the line of fire to rejoin their parent bodies. With equal masses moving at equal speeds in opposite directions, the total momentum before the collision is zero. After the collision, the projectiles have reversed their directions of motion. At what speeds are they now traveling? Momentum will be conserved—that is, it will still be zero—if the two projectiles are moving with equal speeds in opposite directions. And the energy of motion will be conserved in this elastic collision if the speed of the projectiles after the collision is the *same* as it was before the collision. So the velocity of each projectile is simply reversed by the collision.

In the next episode, the duel begins with the opponents farther apart than before. As recorded by a neutral observer, they move toward each other at equal speeds along parallel lines. In accord with the code duello, the duelists fire their projectiles at each other simultaneously. Now, aiming is particularly tricky. Each antagonist fires his projectile broadside, at right angles to his direction of motion. Each projectile moves away on a path that combines this transverse motion with the movement of the duelist who fired it. Both projectiles follow diagonal paths, to collide halfway between the parallel tracks of the two moving duelists. In that collision, the motion of the projectiles parallel to those tracks persists, whereas their transverse motion is reversed. As a result, the two projectiles rejoin their parent bodies at the same time.

Owing to the symmetrical situation of the two moving duelists, energy and momentum are conserved. First, all the

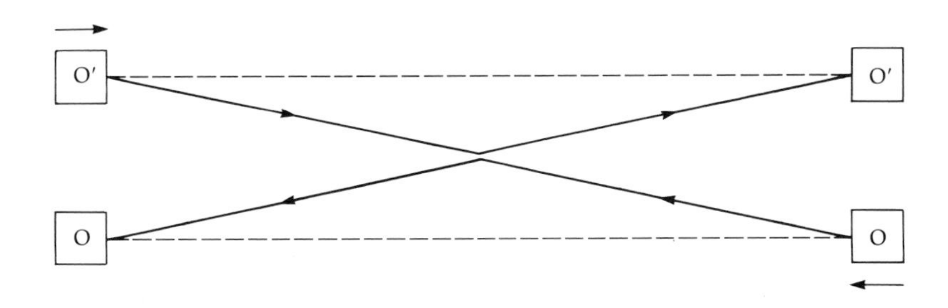

Duelists moving in opposite directions, before and after collision of projectiles

*Why doesn't this ever happen in a showdown at high noon? Although definitely *not* elastic, such a head-on collision does occur in *Logan's Run*,[4] a tale of a future society without a future.

speeds of the two projectiles before the collision and after the collision are the same; energy is conserved. As for momentum measured parallel to the duelists' tracks, each projectile simply maintains, unchanged by the collision, that aspect of its motion; so this component of the total momentum is conserved. Now, let us concentrate on the direction of motion perpendicular to the lines followed by the two combatants, as it might be recorded by the neutral observer stationed between those tracks and following the motion of the projectiles across his line of sight. Along that transverse direction, the projectiles approach one another with equal speed and, after collision, retreat with equal and unchanged speed (as mentioned, their total speeds, which include what is imparted by the motions of the duelists, are also equal). As when both duelists were at rest, the momentum along this direction is zero before and after the collision; it, too, is conserved.

All right, but how can we learn something new? By asking the following question: How does all this appear to one of the combatants—O, for example? He observes his own projectile moving out and returning along the same line, just as when both duelists were at rest relative to each other. But now O′, the other duelist, who is moving rapidly relative to O, fires first!* Foul!

How does it come about that O observes O′ to fire first? Certainly, relativistic dueling has its timing problems—what is simultaneous to the neutral observer will not be simultaneous to O, in relative motion. To understand the particular *sequence* of firings as

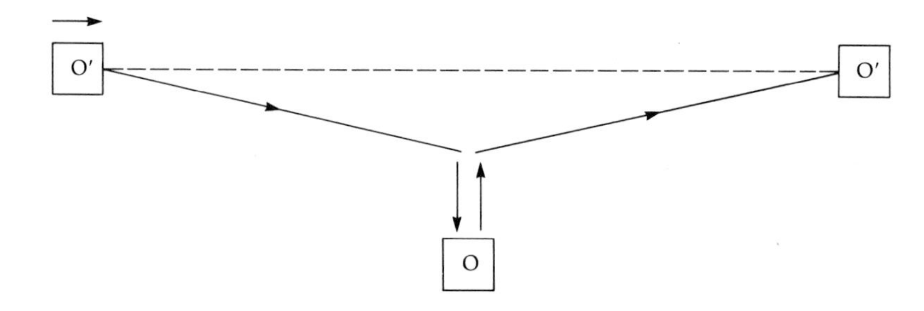

The complete duel as observed by O

*It must not be thought that observer O *sees* his antagonist fire first; the time of which we speak has had removed from it the travel time of the signal that brings to the observer the information of the firing act. If the speed of the moving observer is very close to c, that signal arrives only just ahead of the projectile itself. More will be said later about the distinction between seeing and observing.

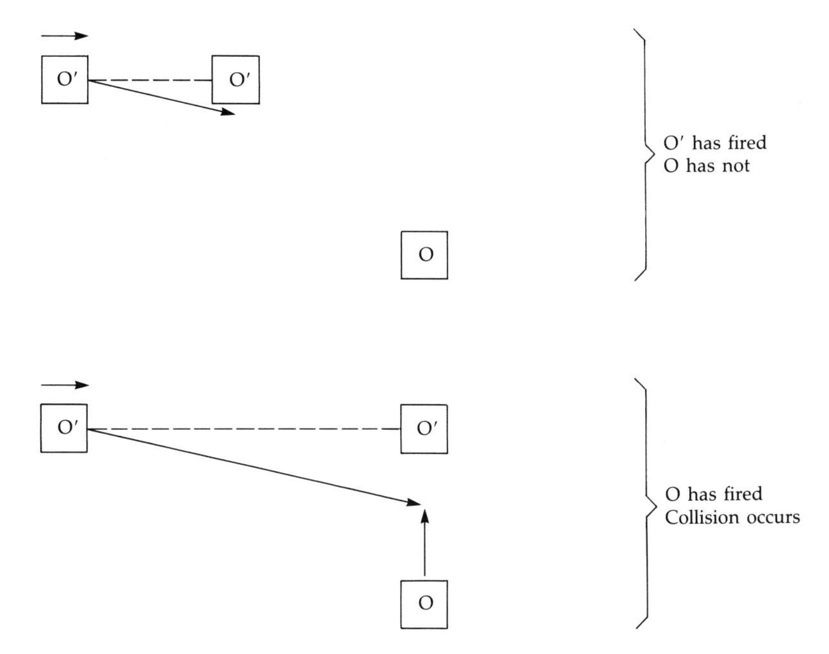

The first stages of the duel as observed by O

observed by O, recall two aspects of relativity developed in Chapter 2, one referring to time, the other to space.

First, relative to observer O, a clock carried by O′ runs slow; the period of the clock is *increased* by the time dilation factor

$$\gamma = \frac{1}{\sqrt{1 - (v/c)^2}},$$

where v is the speed at which O′ is observed to move. This time dilation affects all physical processes associated with O′. Second, all three observers (O, O′, and the neutral observer) agree on distances *perpendicular* to the direction of their relative motion. Therefore, in traveling a common transverse distance, the projectile fired by O′ is observed by O to take a longer time than does O's projectile, longer by the factor γ. In other words, as far as transverse motion is concerned, O′'s projectile is slower than O's; its speed is reduced by the factor $1/\gamma$. To present this in symbols, we write the transverse speeds of the respective projectiles (relative to O) as u (for O) and u' (for O′), so that

$$u' = u/\gamma = u\sqrt{1 - (v/c)^2}.$$

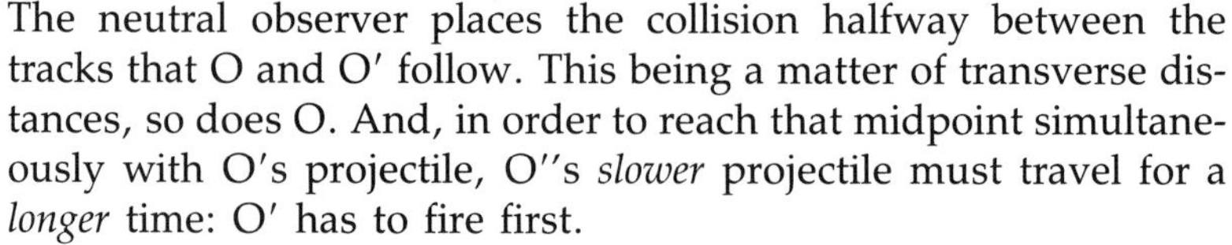

The neutral observer places the collision halfway between the tracks that O and O′ follow. This being a matter of transverse distances, so does O. And, in order to reach that midpoint simultaneously with O's projectile, O′'s *slower* projectile must travel for a *longer* time: O′ has to fire first.

How does conservation of momentum in the transverse direction work now? It is still true that the collision—halfway between the tracks of the observers—reverses the transverse velocity of each projectile. Whatever the total transverse momentum was before the collision, it would be reversed after the collision; that momentum cannot be conserved unless it is *zero*, as it has been up to this point. But how can that be true now? The transverse speeds of the two projectiles are no longer equal—that of O′'s projectile is reduced by the factor $1/\gamma$. Yes, but remember that momentum is the product of *mass* and *velocity*. If the transverse speed of O′'s projectile is reduced relative to O's by the factor $1/\gamma$, the *mass* of O′'s projectile must be *increased* relative to O's by the factor γ, to preserve the equality of the oppositely directed momenta.

Momentum

Velocity

Relativistic Mass

Now suppose that u and u' are *very* small compared with v, and therefore with c. Then the total speed of O′'s projectile is indistinguishable from v, and the speed of O's projectile is effectively zero. Conclusion: The mass of a body moving at speed v (O′'s projectile) is *increased* over the mass of that body at rest (O's projectile) by the factor γ. In symbols, this reads

$$m = \gamma m_0 = \frac{m_0}{\sqrt{1 - (v/c)^2}},$$

in which m_0, the mass of the body at rest, is called, not surprisingly, the rest mass.

Now, at last, we can answer the question posed by the conflict between Newtonian mechanics and the existence of the limiting speed c. It is true that, if a force acts steadily on a body, the momentum of the body increases indefinitely. When the speed is still small compared with c, the mass does not differ significantly from m_0 (i.e., γ is very close to unity), and the increase in momentum produces a proportional increase in speed. But, when the speed has reached a value near c, and can no longer increase very much, the momentum still continues to grow because the *mass* gets larger.

Indeed, it can get larger without limit as v gets closer and closer to c. (See the values of γ in the table on page 54.)

A more intuitive way of expressing this reconciliation comes from the significance of mass as *inertia,* the resistance to acceleration exhibited by a body that is acted on by a force. The increase in inertia as the speed of light is approached implies that the force becomes progressively less effective in accelerating the body, to the point that the speed of light *cannot* be attained. (Although the Force is with us.)

This deviation from Newtonian behavior near the speed of light demands experimental verification. In the early years of this century, the only rapidly moving atomic particles available were beta rays, the fast electrons emitted by some naturally radioactive substances. By measuring the deflection of a beam of beta rays, as produced with controlled electric and magnetic fields, one could study the variation of the electron mass with its speed. The first results, relatively crude, showed that the mass did increase at greater speeds, but they did not favor the Einsteinian prediction expressed by the factor γ. But, by 1910, the weight of evidence had swung in support of Einstein, and, with the later development of accelerators for electrons and other particles, the evidence became overwhelming. We shall return to this topic after taking up the subject of energy.

E IS FOR ENERGY

The principle of *the conservation of energy* emerged slowly. The clear recognition of mechanical energy, and of its conservation, came rather late even though, in January 1669, Christian Huygens had submitted to the Royal Society a paper on elastic collisions that made explicit use of what we would call conservation of momentum and of energy. These ideas were undoubtedly known to Huygens earlier, but on this occasion, which was in the nature of a contest, he had been anticipated a month before by Sir Christopher Wren, now more famous as the architect who rebuilt the churches of London after the great fire of 1666. Yet, throughout the eighteenth century and into the nineteenth century, confusion reigned between the two measures of speed that we call *momentum* and *kinetic energy* (energy of motion). That confusion was further exacerbated by the use of a Latin name for energy that meant *force.*

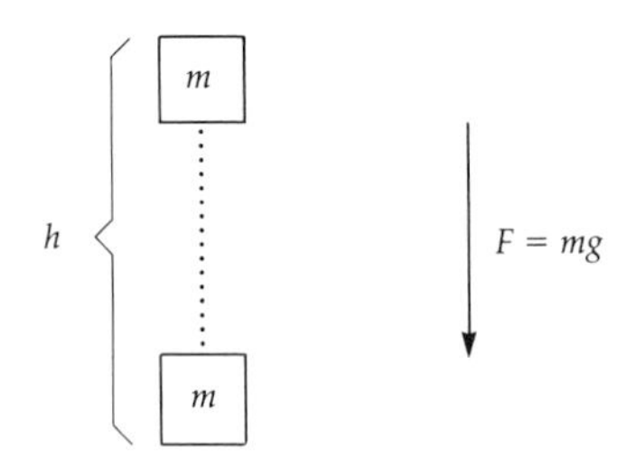

Work

The modern theory of energy begins with the concept of *work;* it is a precise version of the everyday experience of the effort required to move a body against a resisting force, of which the most familiar example is lifting a weight. Lifted bodies of different masses fall back to Earth (paper airplanes are excluded) with a common acceleration, which is called g, the numerical value of g is about 2 percent less than 10 meters per second (speed) per second (acceleration). The force of gravity on a body, which is its *weight,* is the product of its mass m and the acceleration of gravity g, or mg. The *work* performed in lifting a body through a certain height h is measured by the product of the force being opposed (the weight mg) and the distance the body is moved (h). This product, mgh, is the amount of energy that has been transferred to the body by altering its position; that is, by raising it. This energy is potentially available for use when the body is *lowered.* It is *potential energy.*

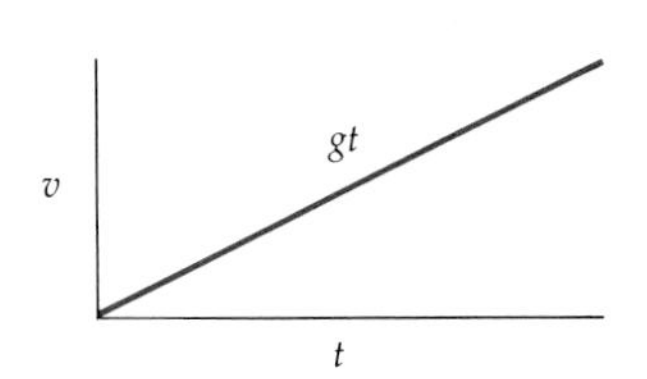

A body falling with the constant acceleration g gains speed at that constant rate. After the passage of time t, its speed (if it began at rest) is $v = gt$. That is the rate at which its height above the ground is decreasing at time t. If this speed were constant, the product of the speed and an elapsed time would be the distance through which the body fell. But the speed is *not* constant; it is zero to begin with and then grows steadily. The average value of the speed is *half* of the final speed. So the distance the body drops in time t is

$$h = \tfrac{1}{2}(gt) \times t,$$

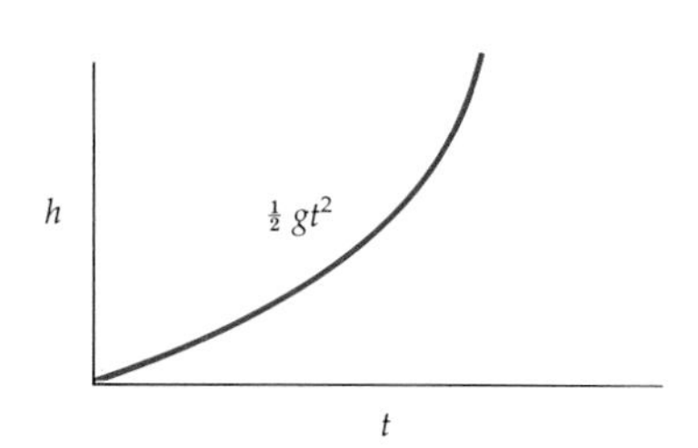

which is $\frac{1}{2}gt^2$. Galileo verified this relation by experimenting with bodies sliding down inclined planes, which reduced the effective force of gravity in a controllable way. This stratagem increased the time of fall from a given height and made it more easily measurable with his crude instruments.*

The work done in lifting the body, the potential energy, returns in the form of energy of motion when the body falls. That energy, which is mgh, is also

$$mgh = \tfrac{1}{2}m(gt) \times (gt) = \tfrac{1}{2}mv^2,$$

*It is very likely that these experiments are the reality behind the legend that Galileo dropped various weights from the Leaning Tower of Pisa.

in which $v = gt$ is the speed that the body acquires when its potential energy is converted into *kinetic energy*. Indeed, $\frac{1}{2}mv^2$ is the Newtonian kinetic energy of a body with mass m that moves at speed v, no matter how that speed is acquired.* For a falling body, potential energy is converted into kinetic energy. Energy flows the other way when a projectile is shot straight up (kinetic energy) and finally comes to a stop at a certain height (potential energy) before starting to fall back to Earth. The swing of the pendulum is a continuous, rhythmic exchange between potential and kinetic energy, the sum of the two remaining constant.

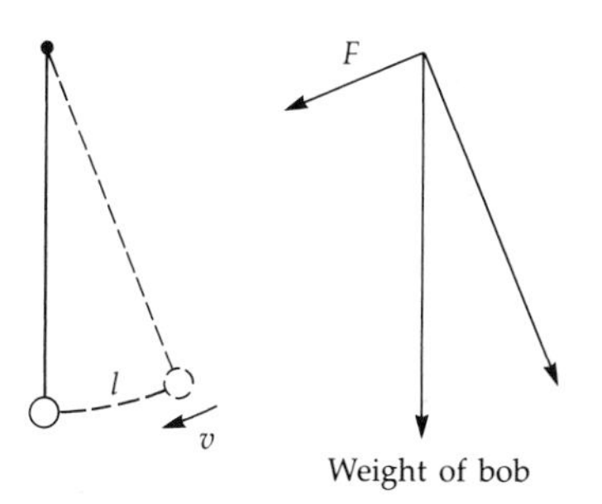

Exchange between potential and kinetic energy

In the example of the falling bob, the downward force of gravity can be resolved into F, *the component along the direction in which the bob happens to be moving at speed* v, *and the perpendicular force component, which is balanced by the tug of the string or rod to which the bob is attached. When the bob travels the small distance* l = vt, *in the short time interval* t, *it transfers potential energy to kinetic energy. That small change in kinetic* energy *equals* Fl, *or* (Ft)v, *in which* Ft *is the corresponding small change in* momentum.

Let us look into the exchange between potential and kinetic energy a bit more. Think of the pendulum bob as it swings downward and transfers potential energy into kinetic energy. As the bob moves a small *distance,* equal to its *speed* multiplied by a short interval of *time,* it gives up potential energy to kinetic energy in the amount of the *force* multiplied by that *distance.* This change of kinetic energy is therefore also the product of three quantities: force, speed, time. But the product of *force* (the rate of change of momentum) with a small *time* interval is the corresponding change of momentum. So the small change in *kinetic energy* is that small change in *momentum* multiplied by the *speed.* This connection between small changes in kinetic energy and in momentum is true generally.

We can recover the Newtonian kinetic energy, $\frac{1}{2}mv^2$, from this connection; it can also be applied to the particle of light, the photon. That particle has a constant speed, c. Hence, any small change of its kinetic energy E is related to a corresponding small change in its momentum p by the constant factor c. On adding up these small changes, we find that the energy E of a photon is related to its momentum p, by

$$E = pc.$$

This is a statement about a particle always traveling at speed c. But Maxwell, who did not view light as particles, already knew

*The quadratic dependence on v of the kinetic energy, the energy of motion, can become a matter of life or death in some circumstances. Stopping a car by applying its brakes is an act of transforming the car's kinetic energy into heat, by means of friction. That generation of heat is proportional to the distance traveled with the brakes applied. Accordingly, the length of unencumbered road required to stop a speeding car varies as the *square* of its speed; a car traveling at 100 km/h requires four times as much distance to stop as it does at 50 km/h.

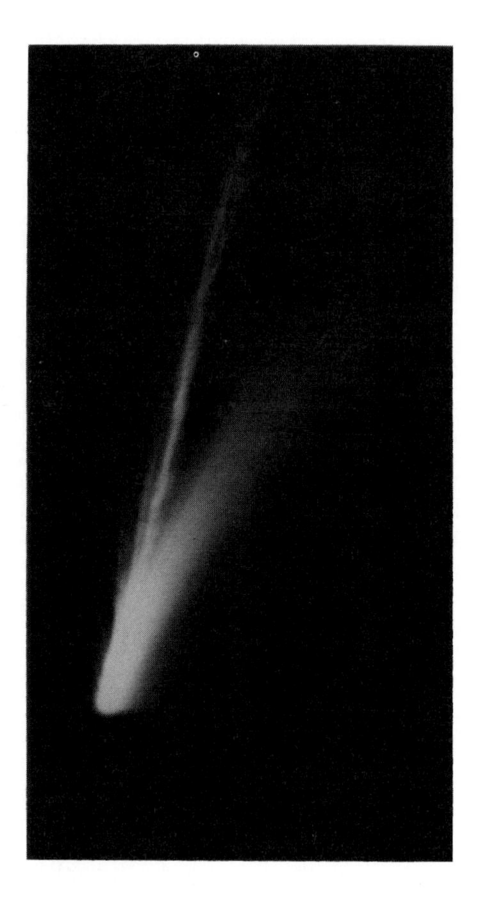

Comet Mrkos, 1957

that a train of electromagnetic waves carries energy and momentum in just this relation. He expressed this idea by saying that electromagnetic waves—light—exert pressure on bodies. We see the momentum of light waves at work in the heavens when a comet, falling toward the Sun, swings through its nearest approach, with a tail that points away from the Sun. Gases and small bits of matter boiled off the head of the comet by the Sun's intense radiation are also pushed away by that radiation, with a force greater than the gravitational attraction of the Sun. (The solar wind, the stream of particles from the Sun, also plays a role here.)

Now notice something. The energy of a photon is $E = pc$, and the momentum of the photon is the product of its mass m and speed c, $p = mc$. Put together, this is

$$E = mc^2, \tag{3.5}$$

and we have it! Yes, but only if E is the energy of motion of a particle traveling at the speed of light. Before we can proclaim the universality of $E = mc^2$, we must widen the scope of E to include other forms of energy. We begin with heat.

Heat

Speculation kept the eighteenth century awash in fluids. There were electric fluids: sometimes two, known as vitreous and resinous; sometimes one (of which Benjamin Franklin was an early advocate). Magnetism had its two fluids, austral and boreal, whereas heat was a solitary fluid, known as *caloric.* A hotter body had more caloric, and it got there by being drained from somewhere else; caloric was conserved. Then along came an adventurer, an American Tory, Benjamin Thomson, later Count Rumford (see Box 3.1).

Experiments that Rumford did in Munich were the basis of *Enquiry concerning the Source of Heat which is excited by Friction,* which was presented to the Royal Society in 1798. In his capacity as Minister of War of Bavaria, Rumford had superintended the boring of cannon. He said, "I was struck by the very considerable degree of heat which a brass gun acquires, in a short time; and with the still more intense heat of the metallic chips separated from it by the borer." He observed that horses, steadily working against the frictional resistance of the metal, caused heat to be produced at a steady rate; the supply of heat seemed inexhaustible. He understood "that any thing which any insulated body, or system of bodies, can continue to furnish without limitation, cannot possibly be a

BOX 3·1 ***Benjamin Thomson, Count Rumford***

Born in Woburn, Massachusetts, in the year 1753, Thomson was sufficiently advanced scientifically by the age of fourteen to accurately predict a solar eclipse. An early marriage brought him to what is now Concord, New Hampshire, then a town contested between Massachusetts and New Hampshire and known respectively as Rumford and Bow. His association with the royal governor of New Hampshire made it expedient for him to leave America when the British evacuated Boston in 1776. Once in London, he rapidly advanced in the affairs of state, while becoming a member of the Royal Society. His multiple pursuits were celebrated by a fellow passenger on a ship to the Continent, one Edward Gibbon, then on his way to Lausanne to complete the *Decline and Fall of the Roman Empire.* Gibbon described Thomson as "Mr. Secretary-Colonel-Admiral-Philosopher Thomson." Having made the acquaintance of the future Elector of Bavaria, Thomson repaired to that state, where for eleven years he was Minister of War, Minister of Police, and Grand Chamberlain and still had time for his scientific activities.

A bit of nostalgia is evident in his choice of title when, in 1791, he was elevated to the nobility of the Holy Roman Empire. Later, he would establish a professorship at Harvard University and the Rumford Medal of the American Academy of Arts and Sciences in Boston.

The origin of the Royal Institution in London (1799) can be traced to an earlier success of Count Rumford in education. The numerous beggars of Munich were a serious social problem. One day Rumford had some twenty-five hundred of them swept up and transported to a previously prepared industrial institute, where they were housed, fed, and taught to support themselves by labor, to the benefit of the electoral treasury. Rumford was operating on a novel principle: "To make vicious and abandoned people happy, it has generally been supposed necessary first to make them virtuous. But why not reverse this order? Why not make them first happy, and then virtuous?"

The present-day inhabitants of Munich are grateful to him for laying out the English Garden, a large park where it is possible to forget that one is in the heart of a city with more than a million people. That gratitude is expressed by a statue, before which a wreath is laid every July 4; American consular officials are conspicuously absent on this occasion.

material substance." The next year, he carefully weighed water as it was frozen into ice and then remelted; these transformations involve substantial amounts of heat. But, although Rumford's measurements were very precise (for the time), he observed no change of weight; caloric was weightless. All this could be understood, however, if heat were a form of "motion of the constituent parts of heated bodies." That in itself was not a new idea,* but here was the first *experimental* evidence.

Did Rumford's work mark the demise of caloric? Not at all. One reason was the failure at that time to distinguish between radiant heat (which *we* know to be electromagnetic radiation) and the heat of material bodies. The concept of heat as a form of internal motion of bodies could easily be rejected because it certainly did not explain the transport of heat through a vacuum. And so, at a time when such distinctions were not made, the faith of the majority of scientists in a substance that is not directly observable was not to be shaken by experiments that, after all, only required that in every substance there is an *unlimited* amount of *weightless* caloric.†

Conservation of Energy

The end of caloric did not come until the 1840s (and even then it lingered on). It took careful experiments to show that mechanical, thermal, electrical, and chemical energies are all interconvertible at fixed ratios. Although there were precursors (in 1842 the German physician J. Robert Mayer wrote: "force [energy] once in existence cannot be annihilated; it can only change its form"), the major credit for establishing the conservation of energy belongs to James Prescott Joule (see Box 3.2). Today, we pay homage to Joule in the name of the unit of mechanical energy, which is the work done, in moving a distance of 1 meter, against a force that would give 1

*Isaac Newton, in his *Opticks*, asked: "Does not . . . light act . . . upon bodies . . . heating them, and putting their parts into a vibrating motion wherein heat consists?"

†The chemists, by this time, had largely abandoned their phlogiston, about which Antoine Laurent Lavoisier (1743–1794) wrote: "chemists have turned phlogiston into a vague principle . . . which consequently adapts itself to all the explanations for which it may be required. Sometimes this principle has weight, and sometimes it has not; sometimes it is free fire and sometimes it is fire combined with the earthy element; sometimes it passes through the pores of vessels, sometimes these are impervious to it. . . . It is a veritable Proteus changing in form at each instant."[5]

BOX 3·2 ***James Prescott Joule***

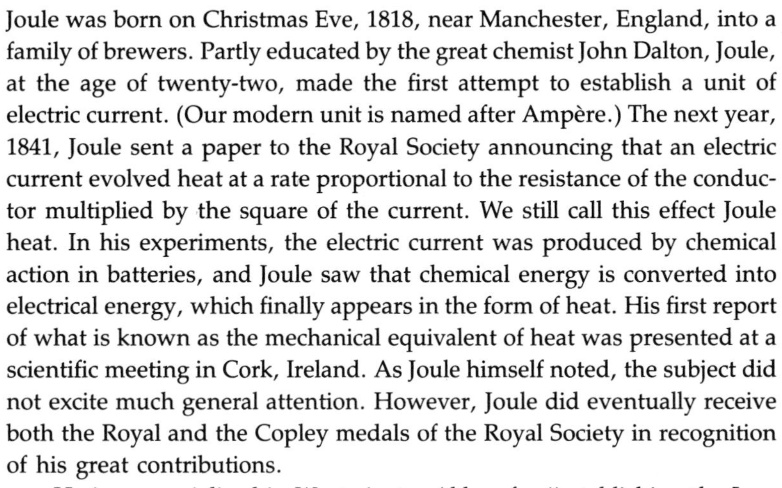

Joule was born on Christmas Eve, 1818, near Manchester, England, into a family of brewers. Partly educated by the great chemist John Dalton, Joule, at the age of twenty-two, made the first attempt to establish a unit of electric current. (Our modern unit is named after Ampère.) The next year, 1841, Joule sent a paper to the Royal Society announcing that an electric current evolved heat at a rate proportional to the resistance of the conductor multiplied by the square of the current. We still call this effect Joule heat. In his experiments, the electric current was produced by chemical action in batteries, and Joule saw that chemical energy is converted into electrical energy, which finally appears in the form of heat. His first report of what is known as the mechanical equivalent of heat was presented at a scientific meeting in Cork, Ireland. As Joule himself noted, the subject did not excite much general attention. However, Joule did eventually receive both the Royal and the Copley medals of the Royal Society in recognition of his great contributions.

He is memorialized in Westminster Abbey for "establishing the Law of the Conservation of Energy and determining the Mechanical Equivalent of Heat."

James Prescott Joule (1818–1889)

kilogram an acceleration of 1 meter per second per second. It is also the electrical energy used when a current of 1 ampere flows for 1 second through a resistance of 1 ohm (named for the discoverer of the proportionality between the current in a conductor and the voltage, or electromotive force). The kilocalorie, or Calorie, the bane of all dieters, which is the heat required to raise the temperature of 1 kilogram of water by 1 degree Celsius, is slightly more than 4,000 joules, or 4 kJ (4.2 kJ is more accurate).

So the idea became accepted that energy can assume many forms but can be neither created nor destroyed—it is conserved. A modern example of the power of the energy-conservation principle had its start in the early 1930s. Experiments on beta rays, the electrons emitted in radioactive decay, showed a loss of energy—only a fraction of the energy available was carried off by the electrons. To save the energy-conservation principle, a new particle, called the neutrino, was invented. It was supposed to be emitted along with the electron and to carry that missing energy, which only

An experimental hall at CERN showing the heavy concrete blocks needed to stop all radiations but neutrinos

apparently disappeared because the neutrino was undetectable. Some twenty-five years later, the neutrino *was* detected and the validity of the conservation principle confirmed. Experiments with neutrinos are now performed rather routinely at high-energy particle laboratories. And suggestions for the technological use of neutrinos are beginning to appear.

Other Conservation Laws

In the 1840s, Michael Faraday recognized experimentally that electricity can be neither created nor destroyed. Material objects are normally in a state of overall electrical neutrality; they exert no significant electric forces on each other. But, if a positive electric charge is produced, whether through rubbing or more elaborate methods, it is always accompanied by an equal negative charge. Paying attention to plus and minus signs, balancing the checkbook of Nature, we find that no net electric charge has been created. And the subsequent recombination of these charges produces complete neutralization; there is no charge left over.

What Faraday could demonstrate with only limited precision is a cornerstone of Maxwell's electromagnetic theory—the theory *demands* the *conservation of electric charge.* The confirmation of this theory by many accurate tests, and the success of its predictions, such as the existence of electromagnetic waves, are the ultimate basis for this conservation law.

The law of *conservation of mass,* first established by Lavoisier, found a place in the atomic theory of chemical reactions initiated by John Dalton at the beginning of the nineteenth century. As an example that uses rounded numbers, 8 kilograms of oxygen (O_2) combines with 1 kilogram of hydrogen (H_2) to form 9 kilograms of water (H_2O). An indirect test of the law of conservation of mass was made in the waning days of caloric. From data for various chemical reactions that require or liberate heat, it was again concluded that caloric was weightless; that is, the *total mass* of the reacting chemicals did *not* change. By 1908, the conservation of mass had been verified to one part in ten million.

Mass and Energy

The convertibility of light energy into other, common forms of energy is a familiar fact of life. Exposure to the rays of the Sun generally raises the temperature of a body; it gains *thermal* energy. If that body is a green leaf, *chemical* energy is produced. The promise of solar power lies in the possibility of direct generation of *electrical* or

A photovoltaic solar power collector

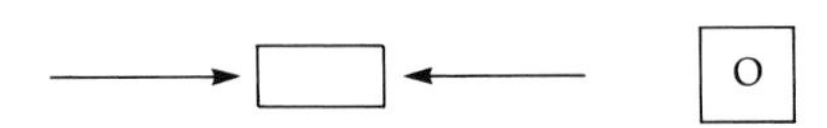

mechanical energy from abundant sunlight. We now proceed to apply what we know about the energy of light, confident that it is typical of *all* forms of energy.

Let us begin with a light-absorbing body, observed at rest. A light beam of definite frequency falls on one side of it and a similar beam strikes it on the opposite side. It suffices to think of two photons, one in each beam. They move in opposite directions and transfer equal energy* to the body. According to relation 3.5, the photons have equal mass, and they have equal speed. Therefore, the photons have equal and opposite momenta, and their total momentum is *zero;* the absorption of the two photons does not change the momentum of the body. At rest before the photons strike it, the body remains at rest after the photons have given up their energy and momentum to it.

Let us say that the photons move horizontally. All four objects with which we deal, the two photons, the body before the act of absorption, and the body after it has absorbed the photons, have zero velocity in the vertical direction (they are either moving horizontally or not at all).

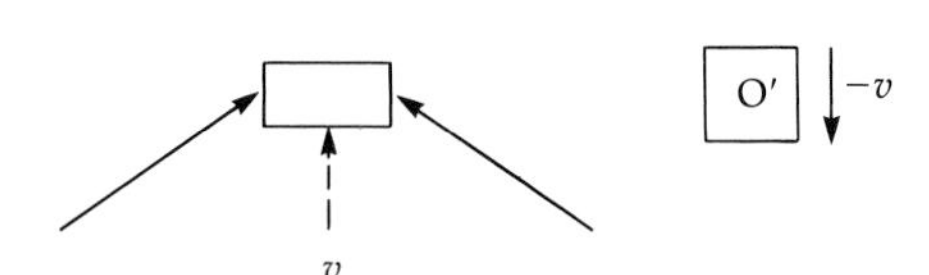

How would this absorption act be described by another observer, who moves with the velocity $-v$ in the vertical direction? Relative to him, all four objects have the common velocity v in the vertical direction. What is this observer's account of the conservation of energy and momentum? We can answer this by using symbols of energy and mass: E and $\bar{E}$ are the body's energies before and after absorption, e is the energy of each photon, m and $\bar{m}$ are the body's masses before and after absorption, and e/c^2 is the mass of each photon.

The conservation of energy tells us that

$$\bar{E} = E + 2e,$$

or that

$$\bar{E} - E = 2e;$$

the energy of the body is increased by the energy of the two photons. The conservation of momentum in the vertical direction is

*Recall that the energy of a photon is determined by its frequency.

this statement:

$$\bar{m}v = mv + 2(e/c^2)v.$$

After canceling the common factor v, we can present this as

$$\bar{m} - m = 2(e/c^2); \tag{3.6}$$

the mass of the body is increased by the mass of the two photons. Explicit reference to the photons can be eliminated:

$$\bar{m} - m = \frac{1}{c^2}(2e) = \frac{1}{c^2}(\bar{E} - E);$$

we have learned that any *increase* in the energy of a body produces a proportional *increase* in mass, with a proportionality factor given by $1/c^2$. Note that we can also consider the emission of photons by the body. Then we find that a *decrease* in the energy of the body produces a proportional *decrease* in its mass. It is usual to indicate a change of either kind by the symbol δ, the Greek lowercase letter delta. Thus, the results we have just arrived at are presented as

$$\delta m = \frac{1}{c^2}\,\delta E,$$

or by

$$\delta E = \delta m c^2;$$

a change in the mass of a body implies a proportional change in its energy, with the proportionality factor c^2.

We know that the mass of a body moving at speed v is related to its rest mass m_0 by

$$m = \frac{m_0}{\sqrt{1 - (v/c)^2}}.$$

In the preceding situation, the speed of the body did *not* change when it absorbed or emitted energy. The change in mass m must be produced by a change in rest mass,

$$\delta m = \frac{\delta m_0}{\sqrt{1 - (v/c)^2}},$$

and

$$\delta E = \frac{\delta m_0 c^2}{\sqrt{1 - (v/c)^2}}.$$

This directs our attention to the opposite situation, and we ask: How does the energy of a moving body change when its rest mass is fixed and its *speed* changes? This is a question about kinetic energy, the energy required to boost the body from rest to some speed or, equivalently, the energy liberated when it is brought to rest from that speed.

Let us begin as before, with a stationary body but replace the two photons with a pair of objects that move at some speed v, less than the speed of light; we call them *particles* so as not to confuse them with the absorbing (or emitting) body. The oppositely moving particles have equal mass and equal speed before they hit the stationary body. Then they come to rest and transfer their equal kinetic energies to the body. This time it is the energy E_{kin} and mass m of the moving particles that interest us, and so let us relabel those properties of the stationary body; we call its energy and mass W and M. Thus, the statement of energy conservation for the body at rest reads

$$\bar{W} - W = 2E_{\text{kin}}$$

Again, we call on the relatively moving observer's account of momentum conservation in the vertical direction. This time, however, we restrict his relative speed (call it u to avoid confusion with the speed of the particles) to be so small that the various masses are virtually the same as with $u = 0$. Thus, the mass after the collision is the sum of the *stationary* body mass $\bar{M}$ and the *rest* masses of the two particles, $2m_0$. On removing the common factor u from the statement of momentum conservation,

$$(\bar{M} + 2m_0)\,u = Mu + 2mu,$$

we get

$$\bar{M} + 2m_0 = M + 2m$$

or

$$\bar{M} - M = 2(m - m_0). \tag{3.7}$$

Then, we apply what we have learned about the mass and energy changes of the stationary body to get

$$\bar{M} - M = \frac{1}{c^2}(\bar{W} - W) = \frac{1}{c^2}(2E_{\text{kin}}) = 2\left(\frac{1}{c^2}\,E_{\text{kin}}\right).$$

A comparison of this with relation 3.7 gives us what we want:

$$m - m_0 = \frac{1}{c^2}\,E_{\text{kin}},$$

or

$$E_{\text{kin}} = mc^2 - m_0c^2. \tag{3.8}$$

Here we see that bringing a particle from rest to some speed v requires an energy change $\delta E = E_{\text{kin}}$ and produces a mass change $\delta m = m - m_0$ such that $\delta m = (1/c^2)\,\delta E$, or $\delta E = \delta mc^2$, as before.

The dependence of kinetic energy on speed v can be made explicit by using the relation $m = \gamma m_0$,

$$\begin{aligned} E_{\text{kin}} &= \gamma m_0c^2 - m_0c^2 \\ &= \left(\frac{1}{\sqrt{1-(v/c)^2}} - 1\right) m_0c^2. \end{aligned}$$

Two things should be said about this result. First, for low speeds (i.e., for small values of v/c) we know that $\gamma - 1$ is accurately represented by $\frac{1}{2}(v/c)^2$. Therefore, we have

$$E_{\text{kin}} = \tfrac{1}{2}\frac{v^2}{c^2}m_0c^2 = \tfrac{1}{2}m_0v^2,$$

which is just what it should be: the Newtonian kinetic energy. Second, for high speeds, as mentioned earlier, γ increases without limit as v/c approaches one. Here is yet another way to understand why it is impossible to accelerate anything to the speed of light; to do so requires what does *not* exist—an *infinite* amount of energy.*

*The photon does not contradict this statement; it always travels at the speed c. But, if we wish to think of a photon as a particle, we must set $m_0 = 0$ to avoid infinite energy. It is not unreasonable that such a particle, which can never be at rest, has no rest mass. Then relation 3.8 tells us that photon energy is entirely kinetic (as did the earlier relation $E = pc$).

Very well, we have shown that $\delta E = \delta mc^2$. How does this differ from $E = mc^2$? They are one and the same if we give δ, the symbol of change, the widest possible meaning. Consider a region of space that is empty of energy and empty of mass; it is a vacuum. Then we introduce into this region a mass m, so that $\delta m = m$. We are thereby introducing an energy E which is $\delta E = \delta mc^2 = mc^2$. In short,

$$E = mc^2.$$

To bring out what is implied by this famous statement, we use relation 3.8 to write

$$E = mc^2 = m_0c^2 + E_{\text{kin}}.$$

Here we see that the total energy of a moving body not only consists in its energy of motion, E_{kin}, but also includes an energy associated with its rest mass, m_0. Yes, but can a significant fraction of this rest energy be released—turned into more accessible forms? The whole world now knows that, in the relativistic realm of atomic particles, the answer is affirmative.

ATOMIC REACTIONS

The rest mass of an atomic particle identifies it—sometimes uniquely but more often as one of a pair of particles that are called particle and antiparticle. As will be discussed later in greater detail, the electron and its equally massive antiparticle, the positron, are distinguished from each other by the sign of the electric charge they carry. That is a frequent but not universal situation. For example, a particle with a mass that is 206.769 times that of the electron is either a positively or a negatively charged μ meson; a particle with a mass of 1,836.152 electron masses (m_e) is either the positively charged proton or its negatively charged antiparticle. However, a particle with a mass of 1,838.683 m_e can be either of two electrically *neutral* particles, the neutron or the antineutron.*

If the rest mass of a particle identifies it either uniquely or as one of a unique pair, then a change in the rest mass must change

*The neutron *is* distinguished from its antiparticle by the sign of another electromagnetic property, the magnetic moment.

the particle. An example is the reaction that signals the decay of the μ meson; that particle disappears and is replaced by an electron (or positron) and a neutrino.* Such reactions are not controllable; they occur spontaneously. Of more general interest are the controllable reactions. Consider, for example, the collision of two particles, called a and b, from which emerges two other particles, c and d:

$$\mathrm{a} + \mathrm{b} \longrightarrow \mathrm{c} + \mathrm{d}.$$

The change in the nature of the particles produces a change in their rest mass, which can be detected by the change in the kinetic energy of the particles.

It is the conservation of energy that lets the alteration in rest-mass energy be sensed through the compensating change in the kinetic energy of the particles. This can be expressed concisely by means of a symbol for the *sum* of things; it is Σ, the Greek uppercase letter sigma. Thus, the sum of the two rest-mass energies before, or after, the reaction will be written Σm_0c^2; the sum of the two kinetic energies before, or after, the reaction will be written ΣE_{kin}. We do not need to specify "before" or "after" in these symbols because we are interested only in the *change* that the reaction produces, and for that purpose we already have the symbol δ. Accordingly, the collision produces a *change* in kinetic energy; a change that we call T,

$$\delta\Sigma E_{\mathrm{kin}} = T,$$

and a *change* in rest-mass energy,

$$\delta\Sigma m_0c^2 = -T. \tag{3.9}$$

Notice that the law of conservation of energy has been obeyed by making these changes equal and opposite. The sum of the two changes is *zero;* the total energy does *not* change.

What is expressed here in symbols is conveyed in words as follows: If, as a result of the collision, the total kinetic energy increases (T is positive), then that energy gain must be supplied through an equal decrease in the total rest-mass energy. Alterna-

*The rest mass of this rather elusive particle is not known accurately—it could be comparable to the electron's mass.

tively, if the total rest-mass energy increases, there must be a compensating decrease in the total kinetic energy (T must be negative). In the latter situation, the reaction cannot take place unless the initial kinetic energy of the particles is larger than $-T$; otherwise, the final kinetic energy would be negative, and no such motion is possible.

Before considering nuclear reactions, let us look at reactions that involve atoms and molecules.

Chemical Reactions

One of the reactions in the burning (oxidation) of hydrogen to form water is the molecular collision

$$H_2 + O_2 \longrightarrow H_2O + O. \tag{3.10}$$

The combination of 1 kg of H_2 with 8 kg of O_2, giving 9 kg of H_2O, liberates a large amount of heat. It is about 30,000 kcal, which would raise the temperature of 300 kg of water from the freezing to the boiling point. This heat is the evidence that energy of motion—kinetic energy—is generated at the atomic level; T is positive in reaction 3.10. Then relation 3.9, divided by c^2, tells us the *loss* of rest mass in the reaction:

$$\delta \Sigma m_0 = -\frac{T}{c^2}. \tag{3.11}$$

How can this be reconciled with the chemical principle of the *conservation* of mass? We must look at *numbers.**

Recall that a kilocalorie is about four kilojoules, or

$$1 \text{ kcal} = 4 \times 10^3 \text{ J (approximately)},$$

and 1 J is the work done in moving a mass of 1 kg a distance of 1 m against a force that would give the body an acceleration of

*As we begin working with very large and very small numbers, it is helpful to use a notation that is based on powers of ten. For example, $100 = 10 \times 10 = 10^2$, $1000 = 10 \times 10 \times 10 = 10^3$, and so forth. Instead of, writing a thousand million, we write 10^9 ($=10^3 \times 10^6$). Similarly, for fractions, beginning with $1/10 = 10^{-1}$, we have $1/100 = 10^{-2}$, $1/1000 = 10^{-3}$, and so forth. Now, if we wish, we can write a microsecond, one millionth of a second, as 10^{-6} s.

1 m/s/s = 1 m/s^2. One joule, then, is

$$1 \text{ J} = (1 \text{ m}) \times (1 \text{ kg}) \times (1 \text{ m/s}^2) = 1 \text{ kg}\,(\text{m/s})^2; \quad (3.12)$$

we recognize here the structure of energy: the product of mass and the square of speed. We now have

$$1 \text{ kcal} = 4 \times 10^3 \text{ kg}\,(\text{m/s})^2,$$

or

$$10^3 \text{ kcal} = 4 \text{ kg}\,(\text{km/s})^2, \quad (3.13)$$

in which the two factors of 10^3 on the right side of the relation have converted meters into kilometers. The speed of light is

$$c = 300{,}000 \text{ km/s} = 3 \times 10^5 \text{ km/s},$$

so that

$$c^2 = 9 \times 10^{10} \text{ (km/s)}^2. \quad (3.14)$$

Relation 3.11 gives the loss of rest mass in a single molecular reaction. Let us add up the losses for all the reactions required to convert 1 kg of H_2 and 8 kg of O_2 into 9 kg of H_2O. That total loss in mass,* δM, is determined by the sum of all the kinetic energies released, which is the total heat generated, H:

$$\delta M = -\frac{H}{c^2}.$$

The amount of heat produced in this reaction is 30,000 kcal = 30×10^3 kcal. Therefore, we have

$$\frac{H}{c^2} = 30 \times \frac{10^3 \text{ kcal}}{c^2} = 30 \times \frac{4 \text{ kg}}{9 \times 10^{10}} = \frac{4}{3} \times 10^{-9} \text{ kg},$$

about a millionth of a gram. Compared with the total mass of 9 kg, the relativistic mass change is roughly one part in ten billion, or

*We can speak here of mass, rather than rest mass, because the additional mass associated with the thermal movement of the molecules at ordinary temperatures is quite negligible.

10^{10}. It has been mentioned that chemical techniques—weighing the reactants before and after a reaction—verify the law of conservation of mass to an accuracy of one part in ten million, or 10^7. At least a thousandfold increase in accuracy would be required to disclose the relativistic mass change. It is in this sense that the relativistic conservation of energy incorporates the chemical law of conservation of mass.

In nuclear reactions, matters are quite different; the scale of energy, and mass change, is a million times as great.

Nuclear Reactions

The nuclei of atoms are composed of protons and neutrons. The proton (symbol: p) is itself the nucleus of the lightest form of hydrogen; the neutron (symbol: n) is an unstable particle. More massive than the proton by 2.531 m_e, the neutron decays, with a half-life of about ten minutes, into a proton, an electron, and a neutrino. The rest mass of this kind of neutrino is known to be very small compared with m_e; the kinetic energy liberated in the reaction is 1.531 $m_e c^2$.

In addition to the lightest form of hydrogen, there are two other varieties, or *isotopes:* deuterium (symbol: D or H^2) and tritium (symbol: T or H^3). The deuteron (symbol: d), the nucleus of deuterium, has the mass 3,670.481 m_e; it is composed of a proton and a neutron. The triton (symbol: t), the nucleus of tritium, has the mass 5,496.918 m_e; it is composed of a proton and two neutrons. All three varieties function chemically as hydrogen because all three nuclei have the same electric charge, that of the proton, and each variety requires only *one* oppositely charged electron to produce a neutral atom.

The mass of the deuteron is less than the sum of the masses of the neutron and the proton,

$$3{,}670.481\ m_e - 1{,}836.152\ m_e - 1{,}838.683\ m_e = -4.354\ m_e,$$

by about 0.1 percent. This difference in mass is revealed in the reciprocal nuclear reactions

$$\mathrm{n} + \mathrm{p} \rightleftarrows \mathrm{d} + \gamma,$$

in which the symbol γ, as used here, stands for a *photon* (γ-ray). When the reaction proceeds to the right, the *decrease* in the rest-mass sum (the photon has zero rest mass), by 4.354 m_0, implies that the kinetic energy of the photon and the deuteron is increased, by 4.354 m_ec^2, over the initial kinetic energy of the neutron and the proton.* When the reaction proceeds to the left, the final kinetic energy of the neutron and the proton is decreased, by 4.354 m_ec^2, below the kinetic energy of the photon and the deuteron. Thus, the latter reaction—the photodisintegration of the deuteron—cannot occur if the kinetic energy of photon and deuteron is less than 4.354 m_ec^2.†

Thus far, the deuteron is just another single particle in a nuclear reaction. Think of it now as a *composite* particle, a neutron and a proton held together by certain forces of attraction—binding forces. There is an energy associated with those forces, E_{bind}, with a corresponding mass, E_{bind}/c^2. Then the deuteron mass, m_{d}, is given in terms of analogously labeled proton and neutron masses by

$$m_{\text{d}} = m_{\text{p}} + m_{\text{n}} + \frac{1}{c^2} E_{\text{bind}}.$$

That tells us the energy of binding of the deuteron,

$$E_{\text{bind}} = -4.354\ m_ec^2.$$

The negative value of this energy of binding means only that at least 4.354 m_ec^2 must be supplied to the deuteron to liberate a proton and neutron.

The mass of the triton also is less than that of its constituents:

$$5{,}496.918\ m_e - 1{,}836.152\ m_e - 2 \times 1{,}838.683\ m_e = -16.600\ m_e,$$

*If the neutron and the proton are moving very slowly—effectively, with zero momentum and zero kinetic energy—the emission of the γ-ray must be accompanied by a recoil of the deuteron in the opposite direction, in order to preserve the zero total momentum. The kinetic energy carried by the deuteron reduces the photon energy slightly below 4.354 m_ec^2; it is 4.351 m_ec^2.

†If the photon and deuteron have zero total momentum, the reaction can begin if the initial kinetic energy is barely greater than 4.354 m_ec^2; the neutron and proton are then produced with negligible momenta and kinetic energies. This is the reverse of the situation considered in the preceding footnote, and so the photon energy at the onset of the reaction is 4.351m_ec^2. But, if the deuteron is at rest before absorbing the photon, the latter's momentum must be transmitted to the neutron and proton. The kinetic energy of these moving particles is also supplied by the photon. Therefore, the photon energy at the onset of the reaction must now be slightly larger: 4.357 m_ec^2.

which gives its energy of binding as

$$E_{\text{bind}}\ (\text{H}^3) = -16.600\ m_e c^2.$$

Unlike the deuteron, the triton is unstable. It decays, with the emission of an electron and a neutrino, into an isotope of helium, He^3; the half-life of the triton is about twelve years.

The most abundant helium isotope, He^4, has two protons and two neutrons in its nucleus; the He^3 nucleus contains two protons and one neutron. The masses of these isotopes are

$$\text{He}^3\text{: } 5{,}495.882\ m_e$$
$$\text{He}^4\text{: } 7{,}294.293\ m_e$$

so their energies of binding are

$$E_{\text{bind}}\ (\text{He}^3) = -15.105\ m_e c^2;$$
$$E_{\text{bind}}\ (\text{He}^4) = -55.375\ m_e c^2.$$

The energies of binding of the three-particle nuclei H^3 and He^3 are similar, He^3 being less strongly bound by 1.495 $m_e c^2$. One can ascribe this difference to the electric energy of *repulsion* between the two protons in He^3, which has no counterpart in H^3. Particularly striking is the increase in the strength of binding in the progression from two to four particles. The resulting fractional decrease in mass is given by the following rough percentages: 0.1, 0.3, 0.8, for the 2-, 3-, and 4-particle systems.

The decay of H^3 into He^3 can be pictured as the decay of one of the neutrons in H^3 into a proton, an electron, and a neutrino. Here, however, the nuclear environment of the decaying neutron plays an important role. In the decay of a solitary neutron, the neutron-proton mass difference, 2.531 m_e, supplies kinetic energy in the amount 1.531 $m_e c^2$. But in H^3, the decaying neutron creates a proton in the presence of the proton that is already there. The repulsion of those protons raises the energy by 1.495 $m_e c^2$, which is subtracted from the available kinetic energy. What remains is*

$$1.531\ m_e c^2 - 1.495\ m_e c^2 = 0.036\ m_e c^2;$$

*The almost complete cancellation of these numbers has motivated the statement of all masses to three figures after the decimal point.

the decay process is barely possible and requires much more time, as observed (the ratio of the H^3 half-life to the neutron half-life is 6×10^5).

This discussion raises a question: If the neutron and H^3 are unstable, why is the deuteron stable? Cannot the neutron in a deuteron change to a proton, forming He^2, with the emission of an electron and a neutrino? Yes, except for one thing; the nucleus He^2 does not exist; the nuclear force of attraction between two protons is not strong enough to produce binding. And so, although 1.531 $m_e c^2$ is made available by the n-p mass difference, at least 4.354 $m_e c^2$ would be needed to destroy the bound state of the deuteron, leaving no positive kinetic energy for the reaction. The instability of the neutron is quenched within the deuteron (and many other nuclei, or you would not be reading this).

The progressive increase in the strength of binding with the increasing number of particles in a nucleus is exploited in the DT reaction,

$$H^2 + H^3 \rightarrow He^4 + n,$$

or

$$(p + n) + (p + 2\ n) \rightarrow (2\ p + 2\ n) + n.$$

There are two protons and three neutrons both initially and finally. Therefore, the change in the sum of the rest masses comes entirely from the energy of binding:

$$\begin{aligned} \delta\Sigma m_0 &= \delta\Sigma \frac{1}{c^2} E_{\text{bind}} \\ &= -55.375\ m_e + 15.105\ m_e + 4.354\ m_e = -35.916\ m_e. \end{aligned}$$

This decrease in rest mass by about 0.4 percent is converted into kinetic energy in the amount of 35.9 $m_e c^2$. That exceeds the energy released in forming one water molecule from its constituents by a factor of nearly 10^7. But perhaps you wonder whether there is some reaction in which *all* the rest mass is converted into kinetic energy? Just this can happen in the mutual annihilation of a particle and its antiparticle.

The Positron

In our day, when it is not unusual for a new particle to be announced at a press conference, it is hard to imagine the situation at

J. Robert Oppenheimer (1904–1967)

the beginning of 1932. Only two basic atomic particles were known, the electron and the proton (and the photon). Then, almost simultaneously, two new atomic particles were found, the deuteron and the neutron. Deuterium was discovered by Harold Urey (1934 Nobelist in Chemistry), the neutron by James Chadwick (1935 Nobelist in Physics). As the deuteron is the simplest nucleus containing a neutron,* it was fitting that the free neutron should make an appearance.

Then came the positron. As it happened, this particle had a place already prepared for it in a recently developed relativistic quantum theory of the electron. Yet this theory seemed so conjectural† that it was not even mentioned in an early report of the discovery. The positron was found in cosmic rays by Carl Anderson (1936 Nobelist), who also played a role (see Chapter 2) in the later discovery of another particle that comes in positive and negative charge varieties, the muon.

Anderson was experimenting with a cloud chamber, a device that shows the tracks of electrically charged high-energy particles by means of little droplets of water that condense along the path of such a particle. The droplets are the visible sign of the electrically charged atomic fragments left in the wake of the rapidly moving charged particle. The cloud chamber, the invention of Charles Thomas Wilson (1927 Nobelist), was the happy product of his early interest in how clouds are formed.

The cloud chamber that Anderson used was immersed in a strong magnetic field. A magnetic field acts on a moving charged particle by deflecting it at right angles both to its motion and to the direction of the magnetic field. Oppositely charged particles swing in opposite directions along circular paths; the signatures of positively and negatively charged particles are unmistakeable. On August 2, 1932, a positively charged particle was seen to penetrate through a thin lead plate inserted into the cloud chamber. Only

*So we now know. It had been supposed that nuclei were composed of protons and electrons (after all, electrons *are* emitted by some unstable nuclei), but this picture had serious difficulties. The neutron itself had to be rescued from the early notion that it was a proton and an electron bound together. We now view the proton and the neutron as two different states of a more basic particle, the nucleon.

†Robert Oppenheimer, who first recognized that the theory required the positron to have the same mass as the electron, later wrote:

"This [production of an electron and a positron] was a theoretical prediction which the theorists who made it were somewhat reluctant to believe until the [positron] was discovered."[6]

Scanning a bubble-chamber picture showing the tracks of positively and negatively charged particles

one positively charged atomic particle was known then—the proton.

Anderson's particle, however, was *not* a proton, which would have stopped a very short distance after penetrating through the lead plate (more about this later). The actual track continued on much farther; it was the track of a *new* particle, of much lower mass than that of the proton. Later studies unveiled the birth and death of the positron. Its creation in cosmic rays is always accompanied by that of an electron—they are created as a *pair* of particles in accord with the conservation of electric charge, the charge of the electron exactly balancing the charge of the positron. The track of the electron is bent into increasingly tight circles by the magnetic field as this charged particle loses energy and straggles to a stop.

The track of the positron is often observed to end abruptly. The positron has met an electron in the water vapor, and they have

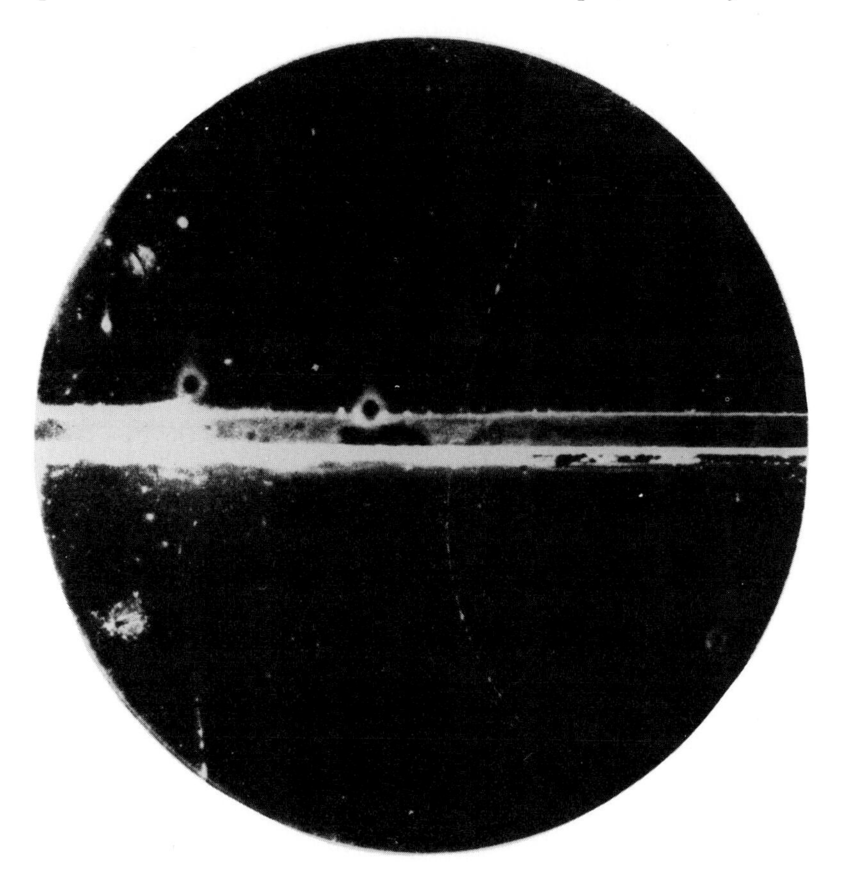

The first published positron track

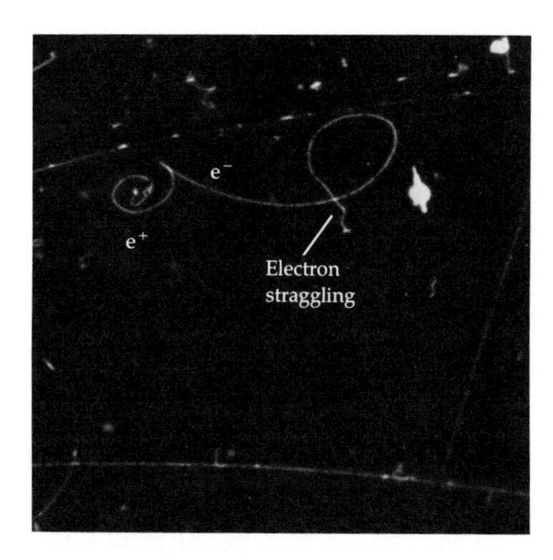

Tracks of an electron and a positron that have been created together

annihilated one another. In their stead, two photons generally appear. Why *two*? A single photon has no electric charge and it can carry away all the energy. But what about momentum? Now, it does occasionally happen that the electron is very near an atomic nucleus, and the recoil of that massive particle will balance the momentum transferred to a single photon. More often, the electron is quite far away from any nucleus, and such a nuclear mechanism cannot operate. Things are particularly simple if both positron and electron are moving quite slowly; the total momentum before the annihilation is essentially zero and must remain so after the annihilation. This requires that two photons emerge with equal and opposite momenta, and therefore with the same energy. They share equally the total energy before the collision; this energy is very close to $m_ec^2 + m_ec^2$, in which m_e is the rest mass of both the positron (symbol: e^+) and the electron (symbol: e^-). It has indeed been observed that, under these circumstances, each photon does have the energy m_ec^2. Thus, the rest masses of the electron and positron are converted completely into the kinetic energy of the photons.

Nuclear Engineering

The mutual annihilation of an electron and a positron is a single atomic event. What would happen if we had a substantial amount of these particles of matter and antimatter, say 0.5 kg of electrons and 0.5 kg of positrons, 1 kg in all, and proceeded to mix them thoroughly? How much energy would be released?

Recall that

$$c^2 = 9 \times 10^{10}\ (\text{km/s})^2,$$

and that a joule is 1 kg $(\text{m/s})^2$. It is more useful here to employ the megajoule, 1 MJ = 10^6 J:

$$1\ \text{MJ} = 1\ \text{kg}\ (\text{km/s})^2,$$

a combination of units that has already appeared, in relation 3.13. Then we find that the energy contained in mass m = 1 kg is

$$\begin{aligned} mc^2 &= [1\ \text{kg}] \times [9 \times 10^{10}\ (\text{km/s})^2] \\ &= 9 \times 10^{10}\ \text{MJ}. \end{aligned}$$

How can we get a feeling for a number like that?

Modern civilization floats precariously on a sea of oil. And oil comes packaged in barrels (bbl), 42 U.S. gallons, or 159 liters. The burning of one barrel of oil produces heat in the amount of 1.5×10^6 kcal $= 6 \times 10^3$ MJ. So the total annihilation of 1 kg of electrons and positrons liberates energy in an amount equivalent to burning the following number of barrels of oil:

$$\frac{9 \times 10^{10}\ \text{MJ}}{6 \times 10^3\ \text{MJ/bbl}} = 15\ \text{million bbl.}$$

As it happens, the current rate of oil consumption by the United States is 15 million barrels per day, a quarter of which has to be imported. As far as energy content is concerned, all that oil flow could be replaced by the total annihilation of one kilogram of mass each day.

Is such total annihilation of matter and antimatter the solution to the impending energy starvation of our civilization? Apart from anything else in the way of technological difficulties, the trouble is that antimatter of any form does not naturally coexist with the matter of Earth. All that energy would have to be pumped in first to *create* the antimatter. If we use naturally occurring materials, it seems that we must be content to convert only a small fraction of rest mass into energy.

Our very existence already depends on such nuclear reactions, which fuel the life-giving rays of the Sun and all other stars. And, in the course of liberating energy, these reactions build up the heavier elements from primordial hydrogen. That *fusion* is the source of all the matter we know, including our own bodies. We are star children.

The *fission* of uranium, discovered in 1938 by Otto Hahn (1944 Nobelist in Chemistry), is a nuclear reaction in which 0.1 percent of the rest mass is made available as kinetic energy. The uranium isotope that undergoes fission, U^{235}, has 92 protons and 143 neutrons, a significant excess of numbers of neutrons over protons (the ratio is 1.55). That excess has the same physical basis as does the mechanism of fission: the strong electric force pushing a proton away from a large number of other protons. As in He^3 and H^3, but on a bigger scale, this electric repulsion reduces the strength of binding of a proton compared with that of a neutron; this tends to favor neutrons over protons in building up heavy nuclei.

The fission of U^{235} begins with the absorption of a neutron to produce the unstable nucleus U^{236}. In one form of this instability, the nucleus undergoes oscillations in which it becomes elongated. Then the strong electric force of repulsion can become dominant over the nuclear binding forces, splitting the nucleus into two fragments, which fly apart at about one-tenth the speed of light; this is nuclear fission.

Let us suppose, for simplicity, that the unstable uranium nucleus splits into equal fragments, each with 46 protons and 72 neutrons; the neutron–proton ratio is 1.56. Such nuclei are isotopes of palladium. But the *stable* isotopes of palladium have about 60 neutrons (the neutron–proton ratio is 1.3). The isotope produced by uranium fission is therefore very unstable. The excess number of neutrons gets reduced in two ways.

First, several neutrons are emitted by the fragments. This is the basis of the uranium chain reaction. One neutron absorbed by U^{235} produces more neutrons, which initiate additional fission reactions, and so on. When the physical circumstances are so arranged that fissions occur at a steady rate (the chain reaction is controlled), one has a nuclear reactor capable of generating great amounts of power.

Second, the fragments that remain after the initial emission of neutrons are highly radioactive. Through a succession of transformations with various half-lives, neutrons are converted into protons with the emission of electrons and neutrinos. And here is the fly in the ointment as far as the large-scale use of fission power is concerned. Some of these radioactive wastes have very long half-lives. Their production in unprecedented quantities poses a serious biological threat to future generations. The promise of unlimited nuclear power seems hollow in the face of this mounting menace. But there may be another way.

We have seen that the DT reaction,

$$D + T \rightarrow He^4 + n,$$

liberates 0.4 percent = 1/250 of the rest mass as kinetic energy. This reaction produces no radioactive wastes (apart from radioactivity that the energetic neutrons might produce in surrounding materials, which to some extent can be controlled). So the energy content of the oil consumed by the United States could be generated by the fusion of deuterium and tritium, in the amount of 250 kg (one quarter of a metric ton) each day, instead of the daily total annihilation of one kilogram of matter and antimatter.

Controlling fusion on Earth—giving engineering reality to these numbers—is quite another matter, however. The deuteron and the triton have the same positive electric charge and therefore *repel* each other. The DT reaction cannot begin until the atoms are given enough energy to overcome that electric repulsion and bring the nuclei nearly into contact.* This requires heating the ingredients of the reaction to the ignition temperature, which is about 200 million degrees Celsius.† So, first, you put energy in. Then you must keep the reacting materials together at a high density for a time long enough that the fusion reaction returns a large dividend on the initial investment.‡ Where nature employs gravitation to hold the reacting matter of a star together, we must make clever use of electromagnetic forces to achieve that goal. The quest for the grail of controlled fusion began more than thirty years ago and is now being pursued in several technologically advanced nations.

*The wave aspect, which all atomic particles exhibit, is basic here. It permits the nuclei to react at energies that, in Newtonian mechanics, would be too low.

†At this temperature a particle has a kinetic energy of about $(1/20)m_ec^2$, on the average. For the deuteron, that implies a speed of about $(1/200)c$.

‡The critical value that must be exceeded is a product of nuclear density ($1/\text{cm}^3$) and confinement time (s) equal to 6×10^{13}, as in 10^{15} nuclei per cubic centimeter held in place for 6×10^{-2} seconds.

(By the end of 1983, both the ignition temperature and the critical product of density and confinement time had been reached—but not in the same machine.)

THE REALITY OF RELATIVITY

It has become clear that the increase of mass with speed that was predicted by Einstein in 1905 is a consequence of $E = mc^2$; the energy that is given to a body to increase its speed also increases its mass. As mentioned earlier, the evidence in favor of Einstein became overwhelming with the development of high-energy particle accelerators. Why is that?

Accelerators

Most particle accelerators are essentially circular in shape. They use magnetic fields to confine charged particles (such as electrons, positrons, or protons) within narrow evacuated tubes while energy is pumped into the particles. We know that a magnetic field forces a moving charged particle to follow a circular path. What determines the radius of the circle? First, the stronger the magnetic field, the smaller the circle. Second, the larger the momentum of the particle, the more difficult it is to bend its path, and so the bigger the radius of the circle. Indeed, the radius is proportional to

The Cosmotron (left), a proton accelerator at Brookhaven National Laboratory, and the tunnel of an accelerator at Fermilab (right)

The Cosmotron was not in operation when this picture was taken. Ordinarily, it is surrounded by heavy concrete shielding blocks. The protons at Fermilab are more than one hundred times as energetic as those of the Cosmotron.

BOX 3·3 *Relativistic Momentum and Energy*

We know that energy $E = mc^2$, momentum $p = mv = (1/c)(mc^2)(v/c)$, and mass $m = m_0/\sqrt{1 - (v/c)^2}$. Therefore, we have

$$E^2 - (pc)^2 = (mc^2)^2 [1 - (v/c)^2]$$
$$= (m_0c^2)^2.$$

This can be used to give the energy for a given momentum,

$$E = \sqrt{(pc)^2 + (m_0c^2)^2},$$

or the momentum for a given energy,

$$p = \frac{1}{c}\sqrt{E^2 - (m_0c^2)^2}.$$

Notice again that, for a photon, whose energy $E = pc$, the *rest* mass m_0 must be zero. But, in general, if the total energy E or the kinetic energy $E - m_0c^2$ is large compared with m_0c^2, either of those energies is virtually equal to pc.

---

the momentum and inversely proportional to the strength of the magnetic field. Put another way, the momentum of the particle is measured by the product of two quantities: the radius of the circular orbit and the strength of the magnetic field.

An accelerator is designed to give the particles a certain kinetic energy. That implies a certain momentum p (see Box 3.3). The design engineer, knowing the maximum strength of the magnetic field, must decide the radius of the circle to be formed by the narrow, evacuated tube that the energetic particles run through. Suppose that a Newtonian engineer (N) and an Einsteinian engineer (E) compete for the contract to design an accelerator intended to give the particles a kinetic energy that is *large* compared with m_0c^2 (there are now quite a few such accelerators in various countries).

To N, that kinetic energy is $\frac{1}{2}m_0v^2$, or $\frac{1}{2}pv$, since his version of momentum is m_0v. In designing this high-energy machine, N will plan for a value of v that is large compared with c. Engineer E, however, knows that the kinetic energy at speeds very near c is very close to pc. Thus, Engineer N arrives at his value of p by dividing the specified kinetic energy by $\frac{1}{2}v$ (which is *much* larger than c); Engineer E divides by c. N's value of p, then, is much smaller than E's value, and therefore N designs an accelerator that is quite compact compared with E's. So N's bid is more attractive economically. Alas, his designs never work! The numerous accelerators that E has built perform as expected.

It was just this behavior of a charged particle in a magnetic field that Anderson used to recognize that his positively charged particle could not be a proton. The radius of the circle in which its path was bent by the known magnetic field told him its momentum. A proton with that momentum would have been moving slowly and with so little kinetic energy that it would have stopped in a short distance. That same value of the momentum for a much less massive particle means that it would move at almost the speed of light, with much more energy. Therefore it would penetrate a greater distance, as Anderson found.

Seeing versus Observing

The discussion of the relativistic duel contained a brief allusion to the distinction between the results of *observation,* in which one corrects for the time it takes the light signals of an event to reach the observer, and things that he actually *sees.* The latter are the raw data of experience, but they are not generally suitable for correlating the experiences of different viewers. Einstein pointed that out in 1905, in discussing how a value of time was to be assigned to distant events.

On Earth, stars are seen in essentially every direction (although there is a concentration in the Milky Way, the plane of our Galaxy). How will those stars appear to the Argonauts on their way to Vega, traveling at a speed, v, that is very close to c? Before leaving Earth, an Argonaut faces Vega, now the pole star, and then turns to pick out a specific and very distant star, off in some other direction (perhaps in the southern sky, where the ancient Argo still blazes in the heavens). The light from that star reaches Earth by combining two kinds of motion, one parallel to the line connecting the solar system with Vega, the other perpendicular (transverse) to that line. Now they set off for Vega. To them, the sources of starlight are in rapid motion with speed v, parallel to the line from Vega. At speeds close to c, with values of γ much larger than 1, the most important effect, as in the relativistic duel, is the slowing down of the transverse motion of the emitted projectile, which is now *light.* Our Argonaut viewer will see his special star in a different direction from before, now much closer to Vega. In this extreme aberration of light at relativistic speed, the starry firmament will be strangely distorted, almost all of it clustered around the Argonauts' remote destination, straight ahead of them (see Box 3.4).

BOX 3·4 ***Relativistic Aberration***

There is a simple geometrical way of displaying the relation between the directions in which any star will be seen by two viewers in relative motion at speed v. Begin with a circle of unit radius and a point on its periphery (as shown at the left below). The line drawn to that point from the center of the circle (E) represents the direction of some star (S) as seen from Earth. The horizontal line is the direction of relative motion; the star Vega (V) is on that line. Now inscribe an ellipse, with the horizontal line as major axis and its left focus (A) situated at the distance v/c from the center (as shown in the middle diagram below). The vertical line dropped from S meets the ellipse at point S′. Then the line drawn from focus A to S′ gives the direction of that star as seen by the Argonauts, moving toward Vega at speed v.

What happens as v/c approaches unity is emphasized by the diagram for that limit, $v/c = 1$. The ellipse degenerates into the horizontal line, and, no matter what the direction of S (with the solitary exception of the precise backward direction), the Argonauts will see it exactly ahead of them (as shown at the right below).

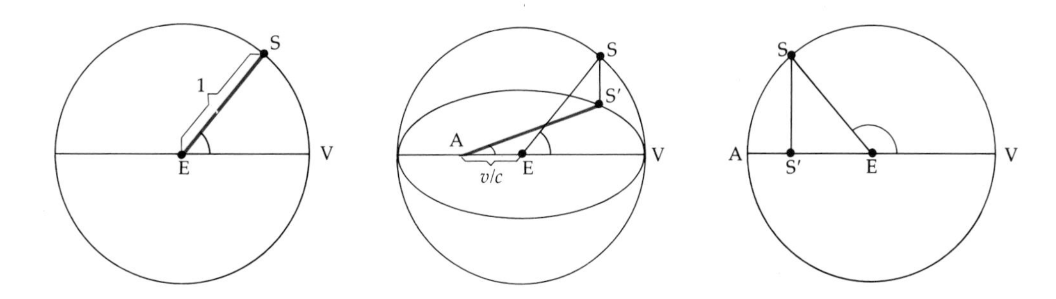

Zeus (Vatican Museum)

Yet any small area of the sky will have the same appearance as before; it is the relation between two such areas in distinctly different directions that is distorted. As a consequence, if their powerful telescope makes out a large planet swinging in a big orbit around Vega, the Argonauts will find it to have the usual spherical planetary shape; it will *not* be seen as contracted along the direction of their flight. There is no contradiction here. The relativistic contraction of length along the direction of motion refers to the *observation* of different spatial points at the *same* time. *Seeing* involves signals received at a common time, which signals originate at distinct points of the viewed object at *different* times.

When the ancient hero Jason and his crew returned from far Colchis, aboard the original Argo, they spoke of the wonders they had seen. What tales the Argonauts of the future will tell! And how differently the star journey will be *seen* by those left behind on Earth. Consider two such eyewitness accounts of the passage of time. Suppose that a beacon is stationed on Earth, high atop Mount Olympus, and an identical one is mounted on the starship Argo before it lifts off. The beacons emit light pulses at a uniform interval, which is the period of these *"clocks."*

The epic journey begins. The once and future Jason, on the Argo, watches the Earth beacon, while Zeus, atop his mountain, looks up at the flashing light in the heavens. On the outward journey, each sees the other rapidly moving away, at speed v. This motion has two effects. Time runs more slowly on the relatively moving body; each successively received pulse is emitted from a greater distance and takes an additional amount of time to reach the viewer. So each sees the other's clock running more slowly. This continues until the Argo reaches Vega. Jason judges that trip to take considerably less time than does Zeus, because, to Jason, the distance traversed is considerably smaller.

The Argo turns around. Now Jason is advancing to meet Zeus's light bolts; each successive one has a shorter distance to travel and arrives sooner, which overcompensates the slowing of time. So Jason *sees* Zeus's clock running faster, all the way home. To Zeus, almost the whole round-trip time has elapsed when he receives the last light pulse emitted by the Argo before it turned about and headed for home. That is because Argo, moving at almost the speed of light, is not far behind! (Recall a similar remark about the relativistic duel.) All the pulses emitted on the return journey, when Jason's clock also is seen as running fast, are received in that short time interval.

Because Jason and Zeus, initially together on Earth, eventually rejoin each other, their *visual* account must contain the same objective lesson about time that the twin observers Castor and Pollux dramatized in Chapter 2. Although each viewer finds the other clock running more slowly on the outbound leg, Zeus has that experience for much longer than does Jason. And, on the return trip, although each sees the other's clock running faster, that episode is of very brief duration for Zeus. In short, Zeus sees the round trip taking much longer than does Jason, and by just the expected factor. Of course, being *Zeus,* he has not aged at all.

Synchrotron Radiation

The phenomenon of synchrotron radiation was mentioned in Chapter 1 as a direct proof of the physical unity of the electromagnetic spectrum, from radio waves to X-rays and γ-rays. Now let us see how it works. Consider an electron moving at low speed (compared with c) in a circular path, as constrained by the magnetic field of an accelerator known as a synchrotron. A moving charged particle is an electric current. This current changes because the moving electron keeps changing direction. The changing current produces as electromagnetic wave, one with the same frequency as that of the circular motion. It is a radio wave, with a wavelength of very many meters. The intensity of the radiation does not vary very much from one direction to another.

More energy is then fed into the electron and it approaches the speed of light; the rate of revolution is much greater. If things continued to work in the same way the frequency of the radiation would be correspondingly increased, the wavelength shortened. Yet it would still be a radio wave, the wavelength being given by the circumference of the electron orbit. But things do *not* work in the same way, for some of the *relativistic* effects just described now come into their own in the real world.

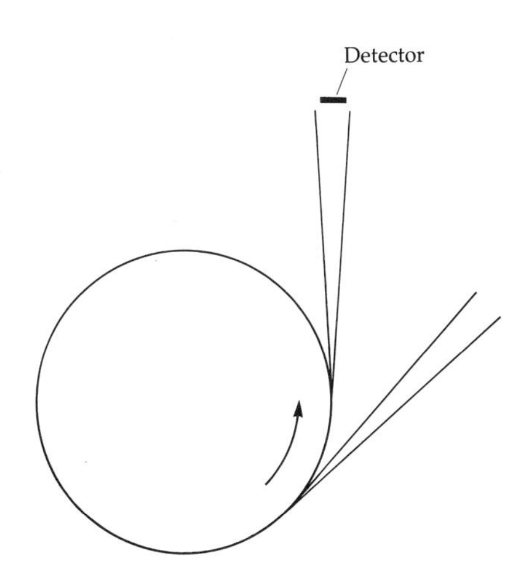

The narrow beam of synchrotron radiation at two points in the electron's orbit

First, there is *relativistic aberration.* The rapidly moving electron emitting radiation is like a rapidly moving star emitting light. And so, the electron's radiation no longer goes off in all directions; it is confined to a narrow beam in the direction of the motion. To detect this beam, a viewer must be almost in the plane of the orbit, and then the only radiation he receives is that emitted when the electron is moving almost directly toward him. In consequence, the detected electromagnetic field changes rapidly as the narrow beam swings past. Rapid change means high frequency; it works out that the frequency is increased by about a factor given by the value of γ

Synchrotron light produced at Brookhaven National Laboratory

for the electron's speed. But that is not all. As in Zeus's view of Jason on the home stretch, the detected radiation arrives in an interval much shorter than the duration of the emission, because the radiating electron almost keeps pace with its own radiation. That produces a further boost in frequency; it is about equal to the *square* of γ.

Taken together these relativistic effects increase the frequency of the detected radiation by a factor that is equal to about the *cube* of γ. The synchrotron at Stanford University that is now routinely used as a radiation source brings electrons up to a value of γ of about 6,000 (6×10^3). The third power of this number is somewhat more than 2×10^{11}! Thus, as the energy of the electrons is raised to the maximum value, the detected radiation changes from radio waves, through infrared, to visible light, on to ultraviolet, and finally into far X-rays. *That* is the reality of relativity.

The space-time world of special relativity is not mysterious. It sets forth the logic of what is physically meaningful, as conditioned by the limited speed with which any physical information can be sent from one location to another. If the phenomena in this world still seem bizarre, that is because our minds instinctively react with the intuition gained from terrestrial experience, where speeds are very small compared with that of light. It may be that life as we know it would be impossible on a planet where speeds comparable to that of light were the norm rather than the exception. But if beings did exist on such a planet, they would have no difficulty with relativistic concepts. They would have had their Maxwell and their Einstein, but no Newton.

NOTES

1. **H. Dukas and B. Hoffmann, eds.:** *Albert Einstein: The Human Side* (Princeton University Press, 1979), p. 5.
2. Ibid., p. 20.
3. **A. Einstein, H. A. Lorentz, H. Minkowski, and H. Weyl:** *The Principle of Relativity: A Collection of Original Memoirs*, English translation, 1923 (Dover), p. 75.
4. **W. Nolan and G. Johnson:** *Logan's Run* (Dial Press, 1969), p. 176.
5. *Encyclopaedia Britannica*, 11th ed., s.v. "chemistry."
6. **H. Boorse and L. Motz, eds.:** *The World of the Atom* (Basic Books, 1966), p. 1214.

4

A MATTER OF GRAVITY

GRAVITY

Aristotle (384–322 B.C.) had this to say about it: "The downward movement of a mass of gold or lead, or of any other body endowed with weight, is quicker in proportion to its size." The decisive rebuttal would not come for almost two thousand years.

In 1636, while under house arrest at Arcetri, Galileo Galilei completed his greatest work, *Dialogues Concerning Two New Sciences,* which was published two years later in protestant Leiden. Let Salviati, Galileo's alter ego, speak:

> Aristotle says that an iron ball of one hundred pounds falling from a height of one hundred cubits* reaches the ground before a one-pound ball has fallen a single cubit. I say that they arrive at the same time. You find, on making the experiment, that the larger outstrips the smaller by two finger-breadths; that is, when the larger has reached the ground, the other is short of it by two finger-breadths; now, you would not hide behind these two fingers the ninety-nine cubits of Aristotle, nor would you mention my small error and at the same time pass over in silence his very large one. . . .
>
> The variation of speed in air between balls of gold, lead, copper, porphyry, and other heavy materials is so slight that in a fall of 100 cubits a ball of gold would surely not outstrip one of copper by as much as four

*The cubit is an anthropocentrically defined unit of length, whose precise size has varied from one time and place to another. The word derives from the Latin for elbow; it is the distance from the elbow to the tip of the middle finger, about half a meter. The height of the leaning tower of Pisa, nearly 55 meters, is not incompatible with the legend that it was the site of the quoted fall of 100 cubits.

David Scott on the Moon

Aristotle (Naples Museum)

fingers. Having observed this I came to the conclusion that in a medium totally devoid of resistance all bodies would fall with the same speed.

Aristotle's physics was in conformity with the limited experience of his age; on land, all motion takes place against resistance. Therefore, greater speed requires greater force or a less-resisting medium. The absence of any medium—a vacuum—implies infinite speed and is therefore inconceivable. It is quite another matter that this tentative beginning became frozen into rigid dogma. It took a Galileo to do two things: defy authority ("In questions of science the authority of a thousand is not worth the humble reasoning of a single individual") and make the great conceptual leap to an idealized situation in which the absence of a resistive medium simplified the behavior of falling bodies.

A year after the death of Galileo in 1642, his pupil Evangelista Torricelli created a vacuum. Before Galileo's time, the action of a suction pump in lifting water had been ascribed to the Aristotelian principle that Nature abhors a vacuum. Galileo noticed that the abhorrence seemed to stop when the water reached a height of about 10 meters, no further elevation being possible. It was left to Torricelli to understand the role of the atmosphere in supplying the pressure that supported a column of water up to this maximum height. To prove this, he reasoned that mercury, a liquid 13.5 times as dense as water, should be supported to a height of only 10 meters divided by 13.5, or three-quarters of a meter. A long glass tube, which was sealed at one end, was filled with mercury and then inverted, the open end being immersed in a basin of mercury. The level of the mercury in the tube immediately sank to the predicted height of three-quarters of a meter above the mercury surface in the basin. And, where the mercury has been, in the space between the sealed end and the mercury level in the inverted tube, was a vacuum. (The responsiveness of this arrangement to atmospheric pressure is the basis of the barometer.)

The development of the airpump thirteen years later made it possible to evacuate larger volumes and to show by direct experiment that light bodies and heavy ones do fall together in a vacuum. And so, it was less a scientific experiment than a bit of show biz when, in 1971, astronaut David Scott stood on the surface of the Moon and said to his audience on Earth,

Well, in my left hand I have a feather. In my right hand a hammer. I guess one of the reasons we got here today was because of a gentleman named

Galileo a long time ago, who made a rather significant discovery about falling objects in gravity fields, and we thought that where would be a better place to confirm his findings than on the Moon. And so we thought we'd try it here for you. The feather happens to be appropriately a falcon, for our Falcon, and I'll drop the two of them here, and hopefully they'll hit the ground at the same time.

How about that! Mr. Galileo was correct in his findings.

Great theater.

The Roman poet Titus Lucretius Carus (96–55 B.C.) had preserved the atomistic ideas of the Greeks for the awakening Western world. His work, *On the Nature of Things,* had a significant influence on the scientists of the seventeenth century, including Isaac Newton. Here is an excerpt:

If anyone supposes that heavier atoms on a straight course through empty space could outstrip lighter ones and fall on them from above, . . . he is going far astray from the path of truth. The reason why objects falling through water or thin air vary in speed according to their weight is simply that the matter composing water or air cannot obstruct all objects equally, but is forced to give way more speedily to heavier ones. But empty space can offer no resistance to any object. . . . Therefore, through undisturbed vacuum all bodies must travel at equal speed though impelled by unequal weights.[1]

Apparently Aristotle did not have it all his own way in the ancient world. Where did that remarkable insight come from? Perhaps Newton was influenced by it, for certainly it was he who gave definitive form to Galileo's discoveries concerning falling bodies.

MATTER AND MASS

Newton's laws of motion make a clear distinction between *weight* and *mass* as properties of matter. The *mass* of a body is the measure of its *inertia,* its resistance to being accelerated by a force. We emphasize this by calling it *inertial mass. Weight,* on the other hand, is a *force,* the force of gravity pulling the body to Earth. (This definition gets qualified somewhat, a bit later.) You will recall that Newton gave a precise form to the gravitational force between two bodies: it varies inversely as the square of the distance between their centers and is proportional to the product of the values of a property characteristic of the bodies. That property is *gravitational mass.* A particular body, in a given location relative to other bodies, has a certain gravitational force acting on it. If a second body, replacing the first one at the given location, experiences a doubled force, say, its gravitational mass is twice that of the first body. In this way, gravitational masses can be assigned to all bodies, as multiples of some standard gravitational mass. Now, a particular body in a given location relative to other bodies has an *acceleration* proportional to the force on it, which is proportional to its *gravitational mass,* and inversely proportional to its *inertial mass.* If the accelerations of all bodies at that location are the same, as Galileo inferred for an airless Earth, the ratio of gravitational to inertial mass must be a *universal constant,* the same for all types and amounts of matter. Then it is possible, and convenient, to define the gravitational mass as *equal* to the inertial mass. That is the deceptively simple statement of what Lucretius foresaw and Galileo demonstrated (to within a few finger-breadths).

Newton appreciated the need for improving the accuracy of the experiments that stand behind this equality. He devised a new one, using pendulums. As mentioned in Chapter 3, the swing of a pendulum is a continuous exchange between potential energy and kinetic energy. At the top of the swing, where the bob is motionless, the energy is all potential; it equals the work done in lifting the weight of the bob from its lowest position. The potential energy is proportional to that weight and is therefore proportional to the *gravitational* mass of the material composing the bob. At the bottom of the swing, the energy is entirely kinetic; it equals $\frac{1}{2}mv^2$, where m is the *inertial* mass of the bob.

If gravitational and inertial masses are always found in the same ratio (so they can be considered equal), that single mass is a common factor in both energies. Then the mass and composition of a bob have no effect on the pendulum's period, which is completely determined by the length of the pendulum and by g, the acceleration of gravity at that location. Suppose, however, that two bobs had different mass ratios; then the period of a pendulum would depend on which bob was used. For example, consider the bob whose inertial mass was larger in proportion to its gravitational mass. Then the transfer of the gravitational potential energy into the movement of this bob would produce a lower speed; for a given value of $\frac{1}{2}mv^2$, a larger inertial mass m implies a smaller v. Thus, for this bob, the swinging action would be slower, the period

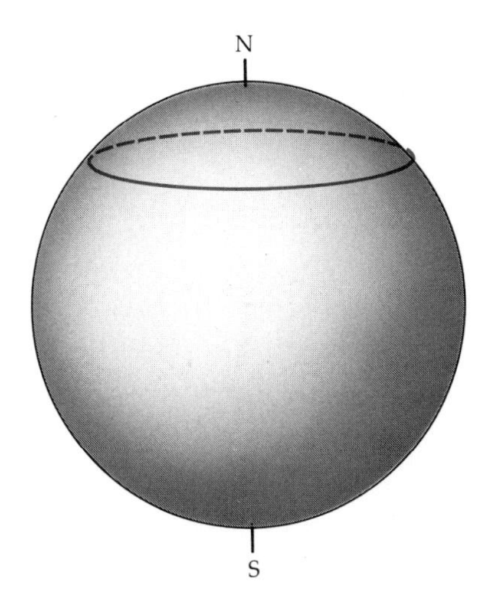

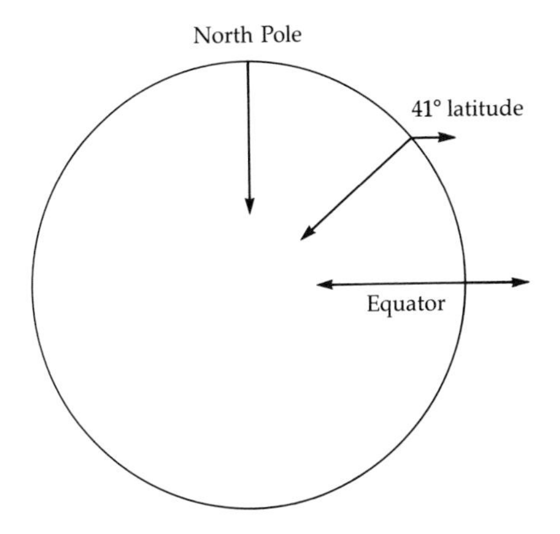

longer. By timing pendulums with bobs constructed of various materials, Newton tried to detect differences in the ratio of the two masses. He found the ratio to be constant, to one part in a thousand. By the first quarter of the twentieth century, the precision of this type of measurement had been increased a hundredfold.

Before the nineteenth century had run its course, however, an experiment of much greater accuracy had been conceived and performed by the Hungarian physicist Baron Roland von Eötvös. To understand it, we must go back to the force that pulls bodies to the Earth, the *rotating* Earth. Among the objections to the Copernican view, with its spinning Earth, was the primitive notion that everything on the surface should fly off, which does not happen. That criticism recognized the inertial tendency of bodies to move in a straight line, but ignored the importance of *relative* motion. During any short interval of time, the motion of any loose object on the surface of the Earth is closely matched by the motion of the Earth beneath it. But, in the course of time, a difference *does* appear; a point on the rotating Earth moves in a circle and therefore falls away from the straight-line path that the loose object is trying to follow. As viewed from the ground, such an object is accelerated *outward* from its center of rotation.

The rotation of the Earth takes a point on the surface through a circular path that lies in a plane perpendicular to the axis of rotation, somewhere between the North and South poles, as determined by the latitude of the point. (For example, both New York and Napoli are nearly at the latitude 41° N.) The acceleration produced by the Earth's rotation, multiplied by the inertial mass of a body, is a *force* that acts on the body in addition to the gravitational force pulling the body toward the center of the Earth. That extra force is absent at the poles, where there is no movement, and is greatest at the equator. There, it acts in the direction opposite that of the gravitational force, producing a slightly weaker *effective* force of gravity.*

*The weaker effective gravitational force at the equator has allowed a relative expansion of that dimension of the Earth; through the ages, the equatorial radius has become about 20 km larger than the polar radius. This larger radius contributes further to the relative weakness of gravity at the equator, where one is farther from the center of the Earth than at the poles (see Box 4.2). As mentioned in Chapter 2, the equatorial bulge is acted on by lunar and solar gravitational forces to produce the very slow change in direction of the Earth's spin axis.

Baron Eötvös recognized the composite nature of the effective force of gravity on a body at the Earth's surface; the effective force combines the force of *attraction* to the Earth's center, which is proportional to the body's *gravitational* mass, with the force of *repulsion* from the center of rotation on the Earth's spin axis, which is proportional to the body's *inertial* mass. Except at the equator and the poles, the combined force does *not* point to the Earth's center. Yet, if the ratio of gravitational and inertial masses is universal—if the two masses can be equated—the direction of the combined force will be the same for all bodies at the same location. If, however, the ratio of gravitational and inertial masses were not the same for all materials, the forces on two such bodies could differ in direction, and this difference might be measurable. But Newton's pendulum experiment had already shown that such a difference in mass ratio could not exceed one part in a thousand; the change in direction would be very slight. How could such small angles be measured?

Eötvös found the decisive tool in the torsion balance, which permits exceedingly minute forces to be sensed, and measured, through their action in twisting a very thin wire or quartz fiber. It had been invented a century previously and first applied in Henry Cavendish's gravitational measurement of the density of the Earth, and in the electrical researches of Charles Augustin Coulomb.

Eötvös's experiment, initially performed in 1889, uses a horizontal beam that is suspended by a fine wire. Two objects of different materials hang from the ends of the beam. The suspending wire balances the total weight. If the ratio of gravitational and inertial masses is not the same for the two materials, and the forces of gravity on the two bodies are not exactly parallel, there will be a net turning force that tends to twist the wire. The sensitivity of this method is so great that the initial experiment demonstrated the absence of any such turning force to a precision a hundred thousand times better than Newton's. With the aid of collaborators, Eötvös continued and improved such experiments, using a variety of substances. By 1922, it had been established that the identity of gravitational and inertial mass held to better than five parts in a billion. Eötvös, however, had already died in 1919, when Hungary was in the grip of the short-lived Soviet government of Béla Kun.

AN ISOLATED FACT

In 1907, Albert Einstein was still at the Swiss patent office in Bern. Not until 1909 would he leave to become what we would call an

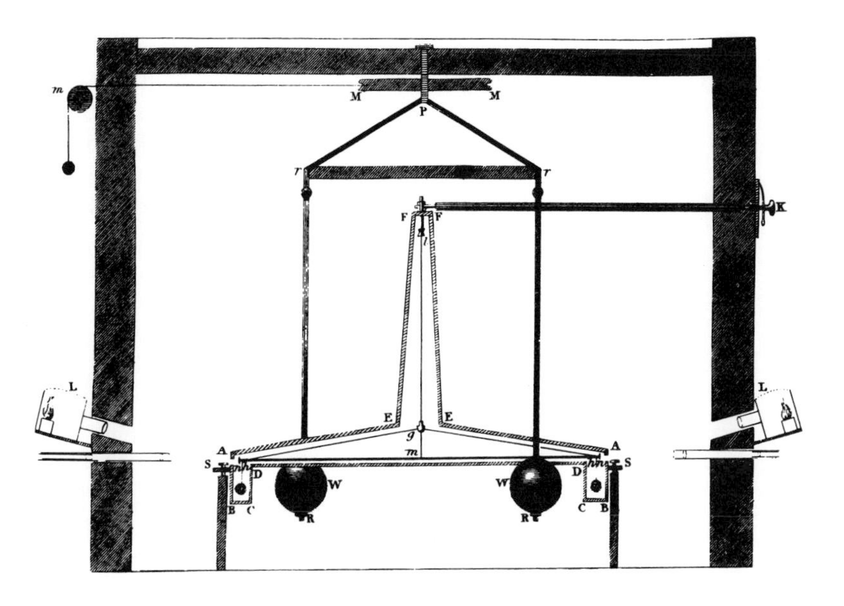

Cavendish's torsion balance

In 1797, he used this apparatus to measure the density of the Earth.

assistant professor, at the University of Zürich. In creating the special, or restricted, theory of relativity, he had dethroned the concept of *absolute velocity*, for in this theory all motion is described *relative* to any *uniformly moving* observer—any observer who is *unaccelerated*. But why should it be required that the observer have no acceleration? Einstein was disturbed by this insistence on the *absoluteness* of acceleration. In a survey paper on relativity, he asks, "Is it conceivable that the principle of relativity also holds for [reference] systems which accelerate relative to each other?"

Starting with this question, Einstein had the sure instinct to focus on the simplest situation, in which the relative acceleration is *constant*. Suppose that several bodies are at rest relative to an inertial observer. How would this look to a second observer who is moving with a constant relative acceleration in a direction that the first observer calls "up"? To the second observer, all the bodies will be moving, with the *same* constant acceleration, in the direction "down". Does that ring a bell? Shades of Galileo! That is just the known behavior of masses falling under a constant force of gravity. Before Einstein, it had been a mysterious regularity of Nature, *an isolated fact*. Now he presented that fact in another way: the equality of gravitational and inertial mass is a consequence of the *equivalence* of a constant gravitational field and a constant acceleration. Concerning the identity of the two masses, Einstein later said,

"The law of the equality of inertial and gravitational masses was now brought home to me in all its significance. . . . I had no serious doubt about its strict validity even without knowing the results of the admirable experiments of Eötvös which—if my memory is right—I only came to know later."[2]

The equivalence of uniform acceleration and a uniform gravitational field was recognized first in the context of simple mechanical motion. But Einstein, seeing in it the answer to the long-standing puzzle of the identity of gravitational and inertial mass, insisted on its general validity. That is the Principle of Equivalence as it later came to be known. This is how he put it in 1907: "We have no reason to suppose that [relatively accelerated reference systems] can be distinguished from each other in any way. We shall therefore assume complete physical equivalence between the gravitational field and the corresponding acceleration of the reference system. This assumption extends the principle of relativity to . . . the uniformly accelerated translational motion of the reference system."

Some years later, Einstein introduced the provocative image of an observer in a windowless elevator trying to decide whether his experiences were produced by a gravitational field or by acceleration. Here is a space-age version.

MISSION EINSTEIN

Prologue

ROBOT GOES HAYWIRE, TEARS SELF APART IN UNIVERSITY LABORATORY

Gainesville, Nov. 27, 1980 (UPI) - An experimental robot ran amok at a University of Florida laboratory, destroying itself before the only person in the University's Center for Intelligent Machines and Robotics Laboratory at the time of the incident last weekend could hit the "kill" switch.

Los Angeles Times

It is the near future. All such early developmental difficulties are thought to have been overcome, for robot astronauts are now routinely employed for difficult space missions. As part of the programming of such robots, the deep-space probe *Einstein* is going to be launched by the space-shuttle orbiter *Challenger.* Owing to a reduction in NASA's budget, the originally planned robot crew of *Einstein* has been cut in half; on board will be only one robot astronaut: Ein.

Kennedy Space Center, Florida. Challenger is positioned on launch pad 39. Attached to it is the liquid-fuel external tank, to which in turn are attached two solid rocket boosters. Lift off! The two boosters are ignited and then separated from the external tank. Before orbital speed is reached, the tank is separated. *Challenger* enters orbit. Relative to Earth, it is now upside down. The doors of the

payload bay open, the remote manipulator deploys *Einstein,* and the probe's rocket engines fire.

Act 1, Scene 1

(Ein is sitting motionless, strapped into a chair before a console, with many switches and lights and a rack of wrenches. Behind the console is a large view screen, now dark. A console light goes on. Ein moves and speaks.)

Ein: The on-board computer activated me. Oh, I see my pre-mission programming, to verbalize everything for the record, was effective. Now, where am I?

(Continually describing his actions, he picks up a wrench and gently parks it in the air, where it hangs. Then he gives the wrench a horizontal tap, and it serenely sails out of the scene.)

Ein: No gravity. I must be out in space, far from any gravitating bodies. Still, let me check that. I'll fire a fast projectile straight ahead and see what happens.

(He pushes a button; there is a slight tremor. He pushes another button and the view screen lights up, displaying the rapidly diminishing red glow of a rocket exhaust, which holds steady in the center of the screen.)

Ein: Same story—no forces are acting on it to pull it down.

(The red glow begins to drift up slowly.)

Ein: Wait! Something is forcing it *up*. Or could it be that something is pulling me *down*? I'll run amok if I don't find out! Where's that down view button? Here it is.

(The button does its work; the visual display changes to a view of the cratered lunar surface rolling by underneath, at a steady distance.)

Ein: I was in a circular orbit around the moon all the time! Then why didn't the gravitational pull make the wrench fall?

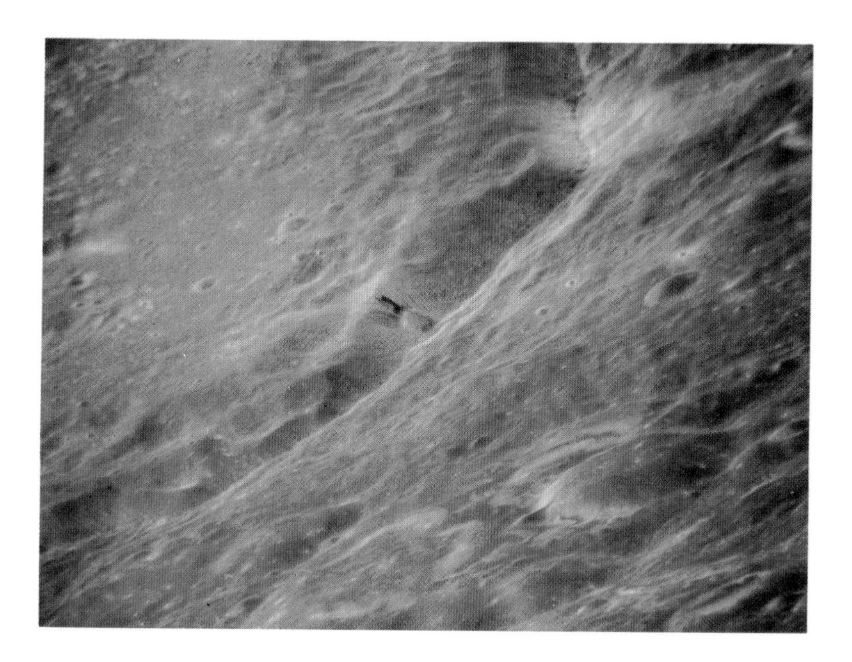

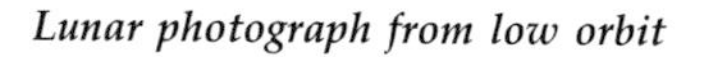

Lunar photograph from low orbit

Gravity and Acceleration

It did fall, but so did the spaceship, and at exactly the same rate. So no *relative* motion was discernible inside the ship. The dimensions of the spaceship are so small compared with the size of the Moon that the gravitational force in the interior, and in the neighborhood of the ship, is very nearly uniform. To an observer on the Moon, the spaceship and all its contents have the same acceleration. They fall together. To another, relatively accelerated observer, one to whom the spaceship is at rest, all the undisturbed objects within the spaceship are also at rest. Thus the equivalence of a constant gravitational field and a constant acceleration implies that the force of gravity can be *cancelled* by a suitable acceleration. That is the situation when the frame of reference is falling freely. You get a taste of that experience in an airplane that suddenly begins its descent: you are somewhat pulled out of your seat; you feel lighter. Complete weightlessness is a routine environment of the space age.*

Spaceship *Einstein*, being in a circular orbit, is doing more than falling freely; it is also moving perpendicularly to the direction of the gravitational force. That motion, by itself, would make the ship fly off in a straight line, and if it did not have such motion the ship would plummet straight down. A proper balance with the gravitational force keeps the ship in orbit: it is continually falling from a straight-line trajectory onto the orbital path. The fast projectile that Ein fired off was moving too rapidly to be held in the same orbit as that of the spaceship.

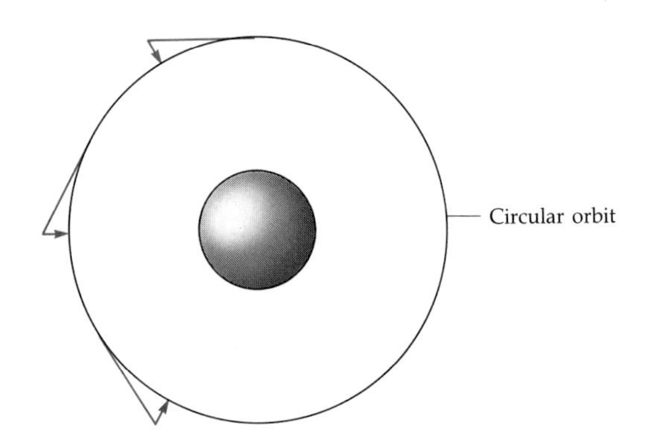

*Adaptation to that environment can have its problems. The story is told of a Skylab astronaut who, after returning to Earth and finishing his first shave, parked the shaving-lotion bottle in the air; there was a splendid crash.

Act 1, Scene 2

(When Computer finished instructing Ein on these matters, she deactivated the robot, took the ship far out of orbit into deep interplanetary space, and set the rocket engines to give Einstein *constantly the same acceleration that lunar gravity produces on the Moon's surface—that is, one-sixth of the corresponding acceleration on Earth, $\frac{1}{6}$ g. Ein has just been reactivated.)*

Ein: Here we go again. But something feels different now. This time I'll use two wrenches of unequal mass. I hold them in the air at the same height, and, at the same time, I LET GO! (*Clang.*) They both fell and hit the deck simultaneously! That's the unmistakable sign of a gravitational field. And, to judge by the weakness of gravity here compared with the Earth, we must have come out of orbit and landed on the Moon. I wonder if they are testing me for the role of David Scott in a rocket Western?

It's a matter of gravity to Ein: $\frac{1}{6}g$. To an unaccelerated observer, it is the deck of the accelerated spaceship that hits both wrenches at the same time. Within the accelerated ship, this is perceived as both wrenches falling to the deck, pulled down by a gravitational force.

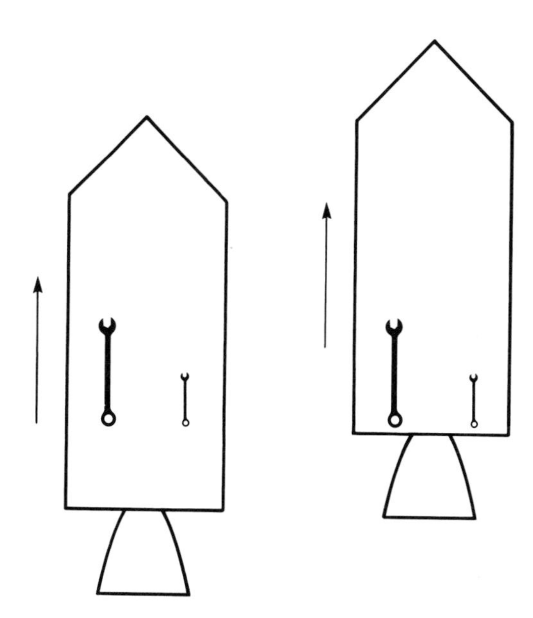

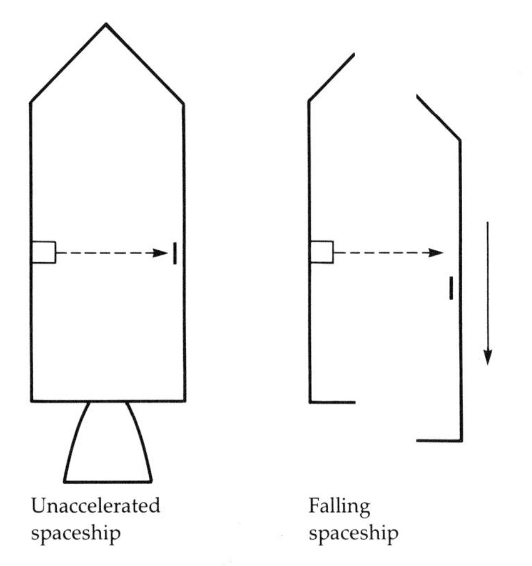

Unaccelerated spaceship

Falling spaceship

Act 2

(Computer has finished explaining that Einstein *is not on the lunar surface but in deep space. Ein, after sitting sullenly for some time, speaks.)*

Ein: That was a nasty trick. For a moment there I had dreams of a little pad in Beverly Hills. But it is particularly annoying that, within the ship, I can't tell when I'm in orbit. It's enough to make one go haywire. So I've thought of a way to do it. Now that the ship is in interplanetary space and can move without acceleration, I will set up a narrow horizontal light beam—a laser beam—and align it to fall on a very small detector. Then, if I am put into orbit again, I should be able to recognize it within the ship. An unaccelerated observer would see the light beam leave the laser, moving in a straight line toward the detector. But, while the beam is traveling, the ship falls in the gravitational field. And so the beam hits the wall *above* the detector, thereby showing the presence of the gravitational field! How about that, Computer?

The Principle of Equivalence

Albert Einstein's Principle of Equivalence states that it is impossible to distinguish—by any physical means whatever—between a uniform gravitational field and a uniform acceleration. Within the spaceship—in orbit about the Moon, for example—gravitation and acceleration have annulled each other, and neither can be recognized separately. To see what Ein, in his state of incomplete programming, has overlooked, let us return to the *identity* of gravitational and inertial masses. As we know, inertial mass and energy are proportional: $E = mc^2$, from which we infer that all types of energy, which produce inertial mass, also produce gravitational mass; *energy,* in whatever form, *gravitates.* That includes electromagnetic energy. Ein's laser beam also must fall in the gravitational field just as the ship does. The beam hits the detector, and all proceeds within the ship as though it were still coasting in interplanetary space, with no sign of either gravitation or acceleration.

This conclusion from the Principle of Equivalence brings up two experimental questions:

1. What evidence is there that various forms of energy do contribute to gravitational mass?

2. Is it possible to show directly that gravitation does influence the motion of light?

Gravitational Energy

In addressing the first question, we shall use the very great precision with which the identity of gravitational and inertial masses has been established: modifications of Eötvös's experiment carried out in the 1960s and 1970s have reached the accuracy of one part in a trillion (10^{12}). First, how about the energy that binds neutrons and protons together in nuclei? Two deuterons, and one helium nucleus (the usual kind), are made of the same particles, two neutrons and two protons. But, as mentioned in Chapter 3, the helium nucleus is much more strongly bound than two deuterons are—in this comparison by an amount of energy that is 0.6 percent of the total mass. We conclude that nuclear binding energy obeys the Principle of Equivalence to one part in $(6 \times 10^{-3}) \times 10^{12} = 6 \times 10^{9}$. What about electrical energy? In the progression from the nucleus of hydrogen, the proton, to the U^{238} nucleus, the most abundant isotope of uranium, which is composed of 92 protons and 146 neutrons, the energy of repulsion between the positively charged particles has grown to 0.4 percent of the total energy. We see that electrical energy obeys the Principle of Equivalence to one part in $(4 \times 10^{-3}) \times 10^{12} = 4 \times 10^{9}$. Can electromagnetic energy be far behind?

A very different test of the Principle of Equivalence probes gravitational energy. To fulfill one of the scientific objectives of the lunar landings, corner-shaped reflectors were left on the Moon; such a reflector sends any electromagnetic signal reaching it back toward its source. Short pulses of laser light were emitted from McDonald Observatory in Texas, and the round-trip times recorded. This gave a precise measurement of the distance from the Earth to the Moon. A series of such measurements was made for a number of years; the accuracy of an individual measurement—over a distance of some 380,000 kilometers—is about half a meter. Now, suppose that the Earth's gravitational mass were different from the Earth's inertial mass. It turns out that such a difference would affect the motion of the Moon in the same way that the gravitational pull of the equatorial bulge of the Earth does. What is actually observed in the long-term effect on the Earth-to-Moon distance is just what that bulge alone should produce. The conclusion from the known experimental errors is that the Earth's gravitational and inertial masses agree to within two parts in 10^{11}. Now, the gravitational energy of the Earth contributes about four parts in 10^{10} to its mass (see Box 4.1), which means that the Principle of Equivalence has been verified for gravitational energy to one part in $(4 \times 10^{-10}) \times (\frac{1}{2} \times 10^{11}) = 20$, which is an accuracy of several percent. Chapter 5 contains a quite different proof that *gravitational* energy *gravitates.*

Gravitating Light

What kind of experiment could show that light does fall in a gravitational field? It certainly helps to have the strongest field available, and for that we go to the surface of the Sun. The mass of the Sun is 333,000 times that of the Earth. The radius of the Sun is 109 times as great as the Earth's radius; the square of this ratio is 11,900. We divide the mass ratio (3.33×10^5) by the square of the radius ratio (1.19×10^4) and find that gravity on the surface of the Sun is 28 times as strong as g, the acceleration of gravity at the Earth's surface. If a light beam were sent horizontally just above the surface of

BOX 4·1 *Gravitational Energy of the Earth*

Given a sphere of matter such as the Earth, a certain amount of energy would be required to dismantle it and spread the pieces far from each other, against the forces of gravitation. Conversely, assembling the whole from its constituents will release that same amount of energy. One way of computing this gravitational energy is to add up the elements of work performed in removing successive thin spherical layers. To make a crude estimate, we can think of a single layer that has one-half of the total mass M, which is moved outward, against the unchanged surface acceleration g, through a distance equal to the Earth's radius R. That work is $(\frac{1}{2})\, MgR$; it falls short, by 20 percent, of the correct answer for a sphere of uniform density: $(\frac{3}{5})\, MgR$. In comparison with the total relativistic energy, Mc^2, the gravitational energy is the fraction $(\frac{3}{5})\, gR/c^2$. Inserting the numerical values $g = 9.8$ m/s^2, $R = 6.4 \times 10^3$ km, $c = 3 \times 10^5$ km/s, we find this fraction to be

$$\frac{3}{5}\frac{gR}{c^2} = 0.6\,\frac{(9.8 \times 10^{-3}\ \text{km/s}^2)(6.4 \times 10^3\ \text{km})}{(3 \times 10^5\ \text{km/s})^2} = 4.2 \times 10^{-10},$$

about four parts in 10 billion.

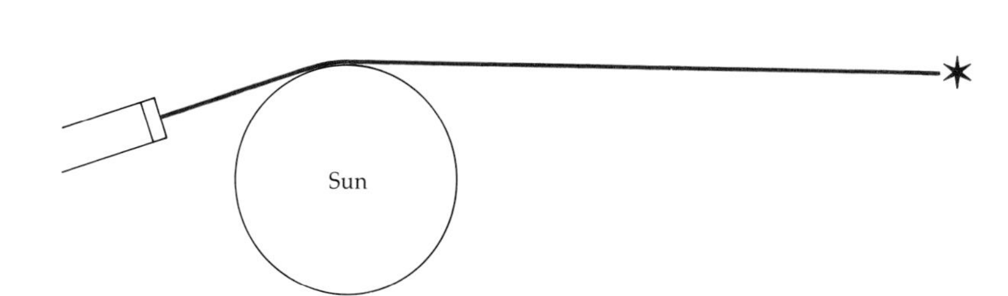

The physical path of light (in color), emitted by a star and observed on Earth, that grazes the sun

The image received on Earth is shown by the black star.

the Sun for a distance of, say, 300 meters, how far would the light fall? The time required for light to travel that distance is one-millionth of a second. Dropping with an acceleration of about 30 *g*, or 300 meters per second per second, it will fall a distance that is all of the diameter of a typical atom.* Quite apart from the uncomfortable circumstances of this experiment, detecting *that* kind of distance would take some doing. (It becomes clear that, even if Ein had been right, his suggestion is quite impractical.) Evidently, much longer distances—distances on the astronomical scale—are required. But then there is no need to do the experiment on the surface of the Sun. We can stay on the cool, green Earth and watch a beam of light, coming from a star, that happens to graze the edge of the Sun. If that light does feel the Sun's tug, its path will be bent, and the star will be seen in a slightly different direction.

We have arrived at just the experiment that Einstein suggested in 1911. He also predicted the size of the effect, and explained how to go about observing it: "As the fixed stars in the parts of the sky near the Sun are visible during total eclipses of the Sun, this consequence of the theory may be compared with experience." What was that "consequence of the theory"? Although Einstein did not put it this way, it might seem to be a straightforward matter. We

*According to Galileo (see Chapter 3), the distance that the light falls is half the product of the acceleration of gravity and the square of the elapsed time, or $\frac{1}{2}(300 \text{ m/s}^2)(10^{-6} \text{ s})^2 = 1.5 \times 10^{-10}$ m $= 1.5 \times 10^{-8}$ cm. The length 10^{-8} cm is a typical atomic dimension.

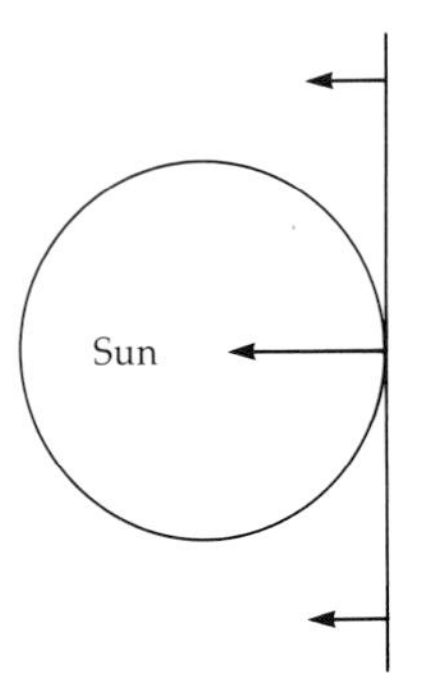

are told that light falls in the same way as any other body does, the only thing distinguishing light being its high speed, c. Indeed, this effect had already been worked out at the beginning of the nineteenth century, with just the result that Einstein found more than a century later. (Einstein was not aware of this anticipation.)

Think of a body moving in a straight line at speed c toward the edge of the Sun. As it passes by the Sun, the force that pulls it at right angles to its initial line of flight first grows, reaches its maximum at the grazing point, and then diminishes. As a result, there is an accumulated transfer of momentum in that perpendicular direction. By comparing this perpendicular momentum with the momentum of the approaching body, we find the angle through which the body is deflected in this close encounter. To simplify things, we recognize that the force transverse to the initial motion is not much different from its maximum value as long as the body is within a solar radius of the grazing point; the force at larger distances will be neglected. Let g' stand for the acceleration of gravity at the solar surface, and m for the mass of the body; the maximum value of the transverse gravitational *force* is mg'. The solar radius is symbolized by R. It takes an amount of *time* given by $2R/c$ for the body to travel the stretch that lies within a distance R of the grazing point. With momentum changing at a rate given by the force, the total amount of *transverse momentum* that is accumulated in that time is $mg' \times 2R/c$. Dividing this by the initial momentum, mc, we get the angle through which the path of the body—light—is bent: $2g'R/c^2$. (This happens to be the right answer to the problem.) This result becomes more significant if we recognize that, in Newtonian mechanics, c^2/R is the acceleration of gravity necessary to hold a body traveling at speed c in an orbit of radius R (see Box 4.2). Thus, the angle of deflection compares the actual acceleration of gravity at the Sun's surface with the *enormous* one necessary to hold light in a grazing orbit about the Sun.

That light-confining acceleration is

$$\frac{c^2}{R} = \frac{(3 \times 10^5 \text{ km/s})^2}{6.96 \times 10^5 \text{ km}} = 1.29 \times 10^5 \text{ km/s}^2 = 1.32 \times 10^7 g,$$

and the deflection angle works out as

$$2\frac{g'}{(c^2/R)} = 2\frac{28}{1.32 \times 10^7} = 4.25 \times 10^{-6}. \tag{4.1}$$

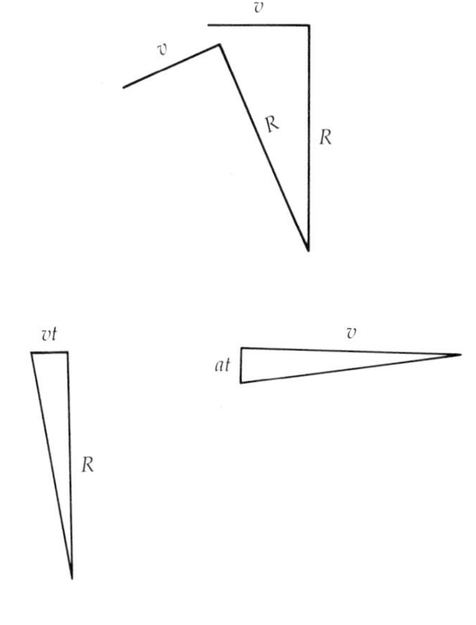

BOX 4·2 ***Rotational Acceleration***

Consider a body that moves in a circle of radius R. At each instant its direction of motion is perpendicular to the line that connects the body to the center of the circle. In a short interval of time, that line changes direction, as determined by the velocity, and the velocity changes direction, as determined by the acceleration. Because the line to the center and the velocity are always perpendicular, these two small changes in direction have the *same* angle. The identity of the angles equates the respective measures of the small changes: the ratio of speed v to radius R, v/R, and the ratio of acceleration a to speed v, a/v. Therefore $a/v = v/R$ and

$$a = \frac{v^2}{R}.$$

This is the inwardly directed acceleration that holds a body with speed v in a circular orbit of radius R.

Gravity

Let us apply this result to the effective weakening of the acceleration of gravity at the rotating Earth's equator. A point on the equator travels the circumferential distance $2\pi R = 2\pi \times 6.4 \times 10^3$ km $= 4.0 \times 10^4$ km in 24 h $= 8.6 \times 10^4$ s, which is the speed $v = 4.0/8.6$ km/s $= 4.6 \times 10^2$ m/s. Thus, the inward acceleration needed to keep a body on the ground at the equator is

$$\frac{v^2}{R} = \frac{(4.6 \times 10^2 \text{ m/s})^2}{6.4 \times 10^6 \text{ m}} = \frac{1}{30} \text{ m/s}^2 = \frac{1}{300}\, g.$$

This is the effective weakening of the acceleration of gravity at the equator of a *spherical* Earth. But Earth is not exactly spherical; the equatorial radius exceeds the polar radius by the fraction $\frac{1}{300}$. That produces a further weakening of about $\frac{1}{600}\, g$ at the equator, so that the net effective reduction at the equator is nearly $\frac{1}{200}\, g$. For dietless weight reduction, travel from a pole to the equator.

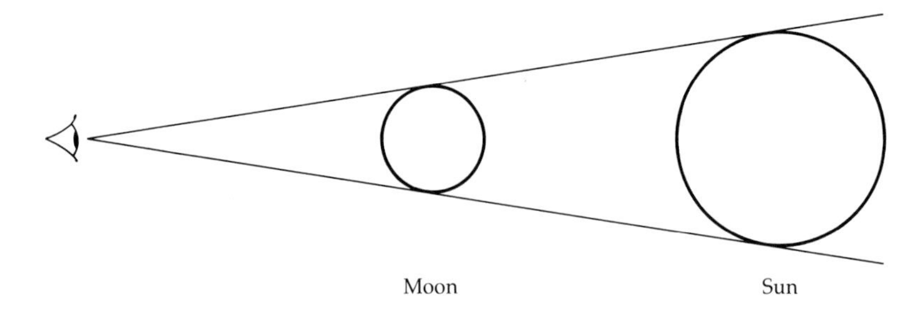

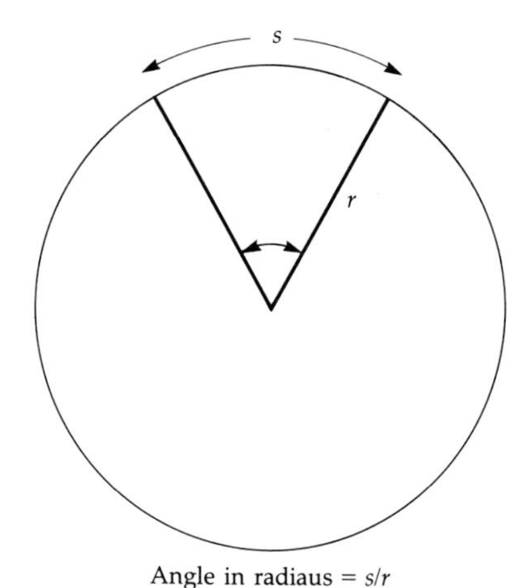

Angle in radiaus = s/r

This is an angle in radians, which is the measure of an angle given by the ratio of the length of a circular arc to its radius. Thus a full circle, 360 degrees, is also a number of radians equal to the circumference of a circle divided by its radius, which is 2π. (Recall that $\pi = 3.14159 \ldots.$) One radian, then, is $360°/2\pi = 57.3°$. It is convenient, in dealing with very small angles, to use seconds of arc; one degree of arc is $60 \times 60 = 3600$ arc sec, and

$$1 \text{ rad} = 57.3 \times 3600 = 2.06 \times 10^5 \text{ arc sec.}$$

And so the predicted deflection angle is

$$(4.25 \times 10^{-6})(2.06 \times 10^5) = 0.875 \text{ arc sec.} \tag{4.2}$$

To get a feeling for this angle, think of the angle that the Moon, and the Sun, subtend at the Earth; it is about half a degree. One sixtieth of that, or half a minute of arc, is 60 km, seen at the distance of the Moon, and half a second of arc is 1 km on the Moon.

Was Einstein right? No, he was only half right. More will be said about this in Chapter 5.

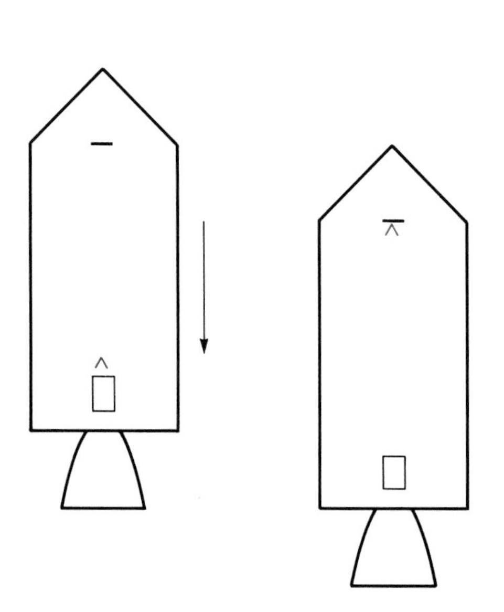

Act 3

(Ein is close to the edge. He snarls at Computer.)

Ein: All right. A horizontal light beam won't work. So I'll try a vertical one! While the light is moving straight up, the orbiting spaceship falls and picks up speed toward the light source. As far as the light is concerned, the source and the detector have a relative motion toward each other, and the Doppler effect comes into play. The detected frequency will be increased, thereby disclosing the presence of a gravitational field. How does *that* compute?

THE RED SHIFT

The Doppler effect to which Ein refers was discussed in Chapter 2. If a detector is moving toward a source of light at speed v, each successive wave is received closer to the source, thereby decreasing the time between successively received crests. The waves arrive more frequently; the frequency is increased. And the measure of that increase, the fractional change in frequency, is the ratio of speed v to the speed of light: v/c. Suppose that the detector in Ein's experiment is mounted at a height h above the light source. The time that it takes the light to travel from source to detector is $t = h/c$. If the spaceship has the gravitational acceleration g' (this symbol was used for the acceleration at the Sun's surface, but now it can have any value), the detector, in the time t that the light takes to travel, will gain a speed toward the source that is $v = g't$. Therefore, the fractional *increase* in the frequency detected is

$$\frac{v}{c} = \frac{g'(h/c)}{c} = \frac{g'h}{c^2}.$$

Once again, however, Einstein's Principle of Equivalence insists that, within the freely falling spaceship, where the acceleration exactly cancels the uniform gravitational field, no sign of either acceleration or gravitation can be found by any physical means whatever. Ein has overlooked once more the effect of gravitation on the light: there must be an exactly compensating *decrease* of the frequency for light moving *up* in the gravitational field. Because red light is the visible light of lowest frequency, any decrease in frequency is called a *red shift*. Before we have Computer tell Ein about this red shift and risk having him go completely haywire, can we come up with a supporting argument for the reality of the gravitational red shift?

Yes. It uses Einstein's photon idea. Recall that the energy of a photon is proportional to the frequency of the light. Suppose that a light source and a light detector are going to be installed at different heights in a spaceship or in a terrestrial laboratory. First the source is positioned. The next step is to raise the detector to a height h above the source. Think of what that means for an individ-

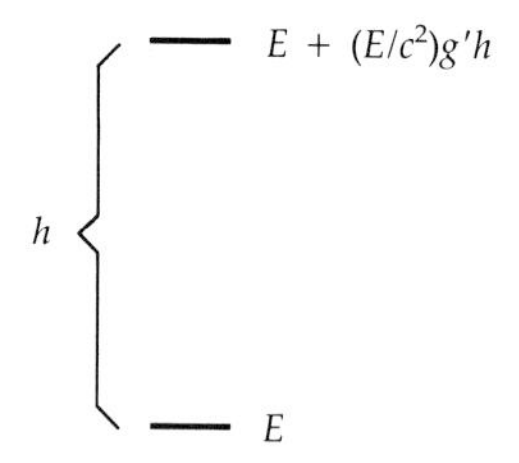

ual atom, one that has some relativistic energy E, at the height of the source. To raise that atom, of mass m, a distance h against the acceleration of gravity, g', requires an amount of work (the product of force and distance) equal to $mg'h$. This work increases the energy of the atom; that is, it has been given additional potential energy. Now, because Einstein has told us that $m = E/c^2$, the additional energy is $(E/c^2)g'h$, which is an increase in energy by the *fraction* $g'h/c^2$. Whatever energy values the atom can have at the height of the source, *all* those energy values are increased by the *same* fraction, $g'h/c^2$, as the atom is raised the distance h.

When an atom emits a photon of light, the energy of the atom decreases and the energy thus made available emerges in the photon. When light is absorbed, things go the other way; the photon's energy is transferred to the atom, increasing its energy. If an atom at the height of the source is capable of emitting or absorbing a photon of a certain energy, which is light of a certain frequency, that same atom, raised a distance h, will emit or absorb light of a *different* frequency, shifted *up* by the fraction $g'h/c^2$. To an atom in the detector, mounted a height h above the source, the light that reached it from the source does *not* have its frequency raised by the appropriate fraction $g'h/c^2$, and the light is received as *down* or *red*-shifted light, shifted by that same fraction $g'h/c^2$. This is the gravitational red shift.

Note that we have taken for granted that the frequency of the light emitted by the source is received by the detector as light of the *same* frequency. One sometimes encounters a different explanation of the gravitational red shift: A photon rising up against a gravitational attraction loses energy and therefore suffers a decrease in frequency. But the energy of the photon, like that of any other projectile (friction aside) is *conserved;* its frequency does *not* change. Rather, it is the *standards* of frequency that differ at different locations.

No law says that the detector must be *above* the source. Suppose that it is below the source. Lowering an individual atom *decreases* its energy, and the frequency of the light absorbed by that atom will be diminished. The light reaching it from the higher source will then be perceived as *blue*-shifted light.

Measuring the Red Shift on Earth

Could such an experiment actually be performed in a terrestrial laboratory? In fact, it *was* done in the 1960s at Harvard University. But isn't the effect of Earth's gravity over an ordinary distance incredibly small? Indeed it is. For example, the value of $g'h/c^2$ for $g' = g = 10\ \text{m/s}^2$ and $h = 22.5$ m is

$$\frac{g'h}{c^2} = \frac{(10\ \text{m/s}^2)(22.5\ \text{m})}{(3 \times 10^8\ \text{m/s})^2} = 2.5 \times 10^{-15}.$$

How on Earth can you measure that?

What is needed is a natural source of radiation with a frequency so well defined that such a tiny shift would be noticeable. A major obstacle is this: when an atom or a nucleus emits a photon that carries a certain momentum, the conservation of momentum demands that the emitting body acquire an equal and opposite momentum—it recoils, as any marksman will appreciate. The energy carried by the recoiling body diminishes the energy of the photon. That is what happens if the emitter is *isolated* in space. In practice, however, the emitter is surrounded by other atoms, which exert forces and absorb some of the recoil momentum in uncontrollable ways. As a result, the photon energy in different emission acts can vary over a range that is much wider than the sought-for red shift, making it quite undetectable.

The technical breakthrough came in 1958 when the German physicist Rudolf Mössbauer (1961 Nobelist) recognized the positive side of these atomic interactions, particularly for the rather rigid structure of a crystalline material. Then, there is a significant possibility that the recoil momentum is taken up, not by one atom or a few atoms, but by the whole crystal! The transfer of the recoil momentum to the enormous mass (on the atomic scale) of the entire material requires an utterly negligible amount of energy—the frequency of the emitted radiation becomes as precise as it *can* be. What is it that ultimately limits the sharpness of a frequency emitted by natural sources?

No atomic or nuclear process goes on forever. When the energy that drives it has gone, the process stops. To measure the frequency of a wave, you count the number of crests in a certain time span. But, as the radiation process continues, it weakens in intensity, just as the strength of a radioactive material diminishes in time; so there must necessarily be some uncertainty about that last, very weak vibration. The measured frequencies unavoidably spread over a range of about one vibration in the half-life of the emitting body. In other words, the range of frequencies is about equal to the inverse half-life; a sharp frequency requires a long-lived emitter. This phenomenon is well-enough understood that

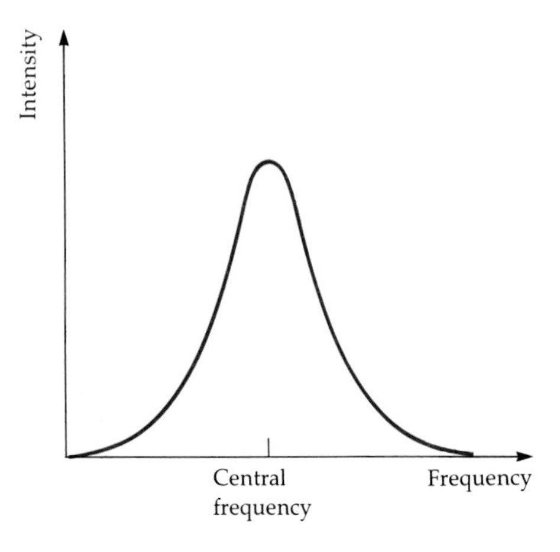

BOX 4·3 ***Spectral Line Shape***

An excited atom is one that has been raised to an energy state above its lowest energy state. Such excited atoms emit radiation in the process of dropping back down to the state of lowest energy. As a result, the number of excited atoms decreases at a steady fractional rate, just as with the radioactive atoms considered in Chapter 2. Recall that the half-life is the time interval during which the number of unstable atoms is reduced by half. A related measure is the mean lifetime, the average time that an atom lasts in the excited state. The mean lifetime is 44 percent longer than the half-life. The variation of the intensity with frequency, which is the shape of the spectral line seen in a spectroscope,* is described by

$$\frac{1}{(4\pi x)^2 + 1},$$

in which x is the deviation of the frequency from that at the center of the line, multiplied by the mean lifetime. The width of the spectral line varies inversely with the mean lifetime. Narrow lines require a long lifetime.

*The first spectroscope was the prism that Newton used to separate white light into its component colors.

the relative intensities of different frequencies within this radiation band also are known (see Box 4.3). Note that a single curve gives the relative intensities when the deviation from the central frequency is measured in units of the inverse mean lifetime.

The need for a very long-lived emitter of radiation was met by using special nuclei. A radioactive isotope of cobalt, Co^{57}, emits a positron and a neutrino to become an isotope of iron, Fe^{57}. This iron isotope is produced, not in its lowest state of energy, but in a state of higher energy, from which—with a very long half-life—it radiates a gamma ray. That half-life is so long that the frequency of the gamma ray can be specified to a precision of one part in 10^{12}. But we are trying to measure an effect a thousand times as small as that. It can be done, by using the known variation of intensity with frequency. To prevent getting bogged down in details, however, let us consider an idealized version of the Harvard experiment, in which the intrinsic spread of the gamma-ray frequencies, and the gravitational red shift being looked for, are comparable.

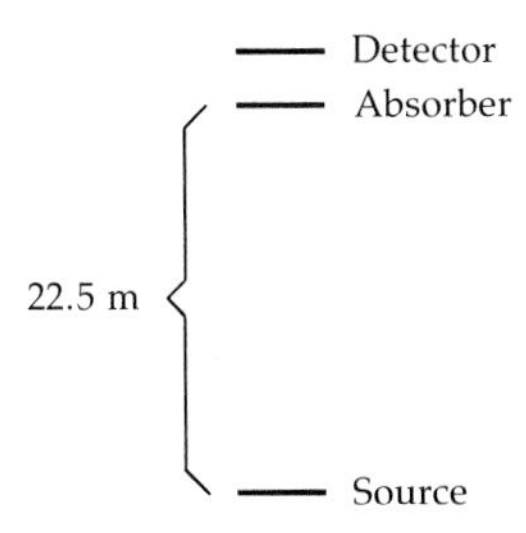

The gamma-ray emitter was positioned at the bottom of a shaft in the Jefferson Physical Laboratory. Several stories above it, at a height of 22.5 m, was the detector, a gamma-ray counter. A sheet of Fe^{57} was placed in front of the detector. That sheet should absorb the gamma rays, in a process that reverses the emission act; the gamma rays impinge on the Fe^{57} nuclei in their lowest state of energy and are absorbed, raising those nuclei back to the state of higher energy. But, if the gravitational red shift is real, the frequency of a gamma ray, after it has risen 22.5 m, should be shifted down enough that it is no longer strongly absorbed. How to prove this? The situation is like that of a radio receiver tuned to a particular broadcasting station. If the frequency of that transmitter should happen to drift, you will lose the signal. To regain it, you must retune the receiver. You can do that here by giving the absorber and detector a motion toward the source; that is, by introducing a Doppler *blue* shift. Properly chosen to compensate the gravitational red shift, just as in the freely falling spaceship, that Doppler shift will restore the state of maximum absorption.

What speed would be required? It is just the speed gained under acceleration g in the time that light takes to go from the emitter to the detector, a distance h. This time is h/c, and the speed is $g(h/c)$, which has the numerical value

$$\frac{gh}{c} = \frac{(9.80\ \text{m/s}^2)(22.5\ \text{m})}{(3 \times 10^8\ \text{m/s})} = 0.735 \times 10^{-6}\ \text{m/s}.$$

At that snail's pace, it would take nearly a year* to close the gap between source and detector; the time required is

$$\frac{22.5\ \text{m}}{0.735 \times 10^{-6}\ \text{m/s}} = 3.06 \times 10^{7}\ \text{s},$$

*This interval of almost a year is independent of the particular height chosen, as can be seen by replacing the specific numbers with symbols; the time required to close the gap h is

$$\frac{h}{(gh/c)} = \frac{c}{g}.$$

In the world of Galileo and Newton, c/g could be interpreted as the time required, under the constant acceleration g, to reach the speed c. But we are not in that world. As was pointed out in Chapter 3, because the speed of light is unattainable, a constant acceleration *cannot* be maintained indefinitely.

whereas the number of seconds in a year is

$$365\tfrac{1}{4} \times 24 \times 60 \times 60 = 3.16 \times 10^7.$$

In the actual experiment, both the red shift (the gamma ray goes up) and the blue shift (the gamma ray comes down) were measured. To eliminate systemic errors that might affect the two measurements oppositely, it was the averaged magnitudes of the two fractional shifts that was compared with the prediction of Einstein's Principle of Equivalence. They agreed to one part in a hundred.

Measuring the Red Shift in the Heavens

Wouldn't astronomical tests of the gravitational red shift be easier? Einstein suggested in 1907 that light from the Sun, observed on Earth, is red shifted. By how much? The fractional red shift of light in rising a *small* distance h, at a point where the acceleration of gravity has the value g', is $g'h/c^2$. We now need the fractional red shift when light travels all the way from the surface of the Sun to the Earth. To get it, we add up the successive contributions from short-path segments, within each of which g' is essentially constant, although g' diminishes as the light moves farther away from the Sun. For a rough estimate of the answer, let us suppose that g' is not significantly different from its value on the solar surface when the distance from the surface is less than R (the Sun's radius) and that it is negligible at greater distances. That gives the fractional red shift as $g'R/c^2$ (this happens to be the right answer), which is just one-half of Einstein's 1911 prediction for the angle through which light, grazing the Sun, is deflected (see relation 4.1). So the solar red shift should be about two parts in a million. This can also be expressed as the relative speed v that would produce an equal Doppler shift. Multiplying the fractional red shift by the speed of light, 3×10^5 km/s, gives v: 0.6 km/s; it is a little less than twice the speed of sound in air at ordinary temperatures. That's not very fast.

There's the rub. In the outer regions of the Sun, heat is transferred by convection. Hot gases move up to the surface, radiate their energy, and fall back into the interior. All that motion quite confuses the search for the small gravitational red shift. Not until the 1960s was a reliable measurement performed. It used one of the well-defined frequencies that are emitted in the yellow light of sodium vapor. This emission occurs in the part of the solar atmos-

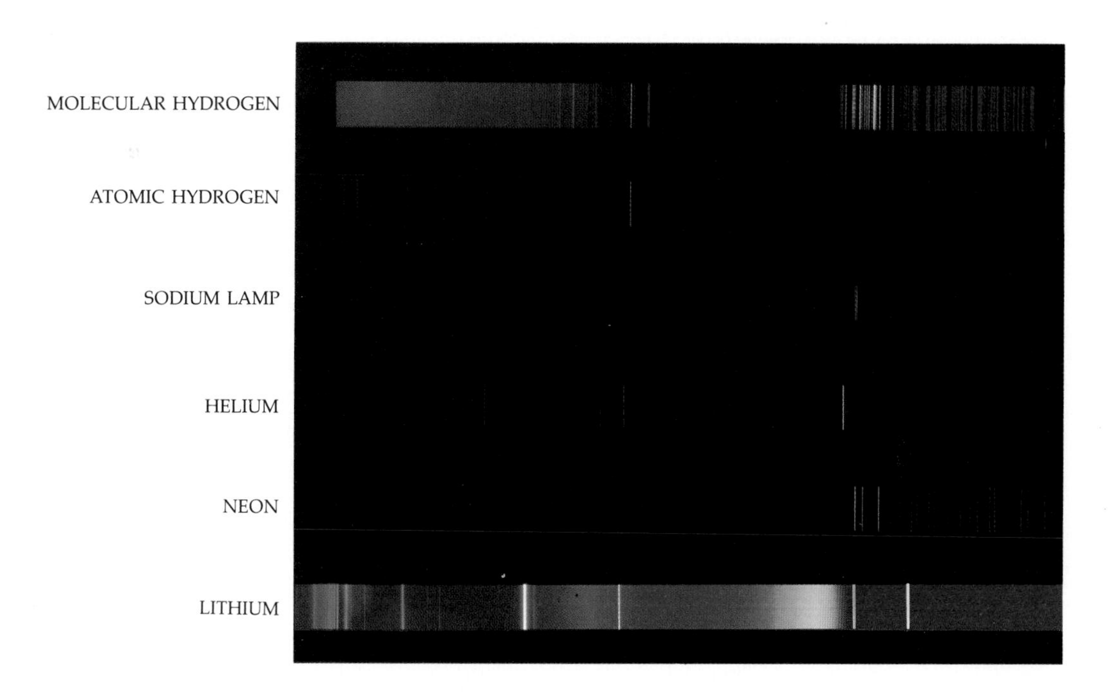

Visible spectra

phere that is comfortably above the region of intense convection, but below the very hot chromosphere. The outcome of those measurements agreed with the expected red shift to within 5 percent.

How about other stars? The gravitational red shift is proportional to the mass and inversely proportional to the radius of the star.* A class of stars called white dwarfs have masses like the Sun's and radii like the Earth's. They are *dense*. Such stars should show much larger gravitational red shifts. To be useful, a white dwarf must have one or more normal companions. Then, by observing a companion, astronomers can learn the mass of the white dwarf and can compensate for the Doppler frequency shift that is produced by the *motion* of the star along the line of sight. Early attention focused on the binary star Sirius, whose white-dwarf component was first seen telescopically in 1862. But a reliable red-

*The fractional red shift is $g'R/c^2$, in which g' is the acceleration of gravity at a distance R from the center of a body of mass M. Newton's gravitational law tells us that g' is proportional to mass M and inversely proportional to the square of the distance R; that is, g' is proportional to M/R^2. Therefore, $g'R$ is proportional to M/R.

shift measurement of a white dwarf was not made until 1954. This measurement used 40 Eridani, an unusual star in the constellation Eridanus, the River. The star is a triplet of two reddish stars and a white dwarf. The gravitational red shift was confirmed to within 20 percent (the difficulty here was pinning down the radius of the white dwarf).

So astronomical measurements of the gravitational red shift cannot compare in accuracy with the Earth-based measurement. But that, in turn, has been surpassed by measurements in near-Earth space. The terrestrial experiment depended on the sharply defined frequency of an electromagnetic oscillation, which Nature provided in the form of gamma rays emitted by the nuclei of atoms in crystalline matter. Now, through the construction of extremely stable oscillators—atomic clocks—mankind has bettered Nature by a factor of a thousand or more.

Atomic Clocks

Since 1967, the international unit of time, the second, has been standardized with the aid of the cesium-beam atomic clock. The atom of the cesium isotope Cs^{133} has a lowest energy state that is split in two by the magnetic interaction between an electron and the nucleus. (This is called hyperfine structure.) The radiation emitted in the transition between these two close energy states has

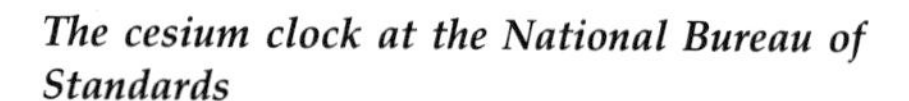

The cesium clock at the National Bureau of Standards

a frequency in the microwave region; it is about 9,193 MHz, which is equivalent to a wavelength of 3.26 cm. The international second is *defined* as 9,192,631,770 periods of that microwave radiation. The instrument that holds the microwave frequency to the necessary great precision makes clever use of magnetic fields to sense whether the radiation in an enclosed region is in exact tune with the atomic vibration, and acts to retune the radiation if it is drifting away.

Such clocks beat at a rate that is the frequency of an atomic radiation. From what we know about the gravitational red shift—the frequency of such a radiation is *increased* if the atom is *raised* in a gravitational field, *decreased* if it is *lowered*—it follows that, when we compare two otherwise identical clocks, the one that is higher in the gravitational field runs more rapidly, the lower clock more slowly.

Atomic clocks have been placed in airplanes and held aloft for many hours to check whether a flying clock does gain time relative to a ground-based clock. In fact, two effects are at work here. The clock moving at speed v runs more *slowly* (special relativistic time dilation) by the fraction $\frac{1}{2}(v/c)^2$, whereas, being at some elevation h, it *gains* time by the fraction gh/c^2. For motion at 70 percent of the speed of sound in air (at ordinary temperatures), 230 m/s, the two effects cancel at a height of 2.7 km. The gravitational effect is larger at the usual cruising altitudes of jet aircraft, about 10 km. The reality of the gravitational effect has been verified in such experiments to an accuracy of 2 percent.*

Einstein understood from the beginning that clocks at rest in a gravitational field run more slowly than clocks outside that field. All physical processes slow down, including the speed of light! Nothing makes it more dramatically clear that the introduction of gravitation is a complete break with the principle of *special* relativity. The extension of the relativity principle to admit *accelerated* frames of reference introduces *gravitation,* and gravitation influences the *speed* of light.

What observable consequence might flow from the slowing of light in a gravitational field, the more so the stronger the field?

*These experiments are complicated somewhat by the rotation of the Earth. Relative to the poles, an airplane flying east has a greater speed than one flying west at the same ground speed. But, for trips around the Earth in opposite directions, averaging the two fractional time changes removes the influence of the Earth's rotation.

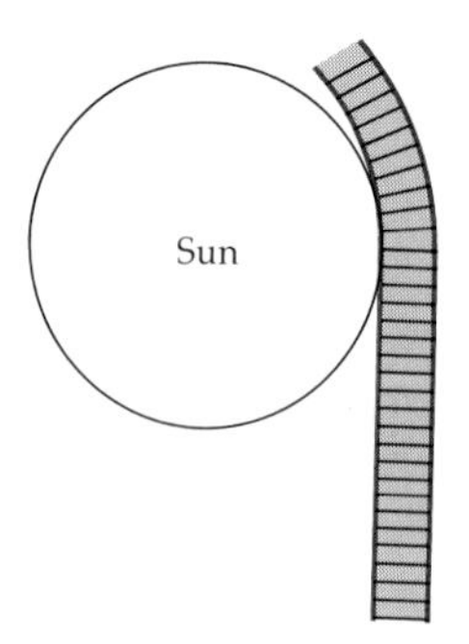

Have you ever watched soldiers or a marching band on parade? Marching abreast they reach an intersection; the marchers on one side slow their pace, while those on the other side maintain it. What happens? The parade turns. Likewise, as a beam of light, of a certain width, passes near the Sun, the part near the Sun travels more slowly than the outer part; the beam turns. It bends toward the Sun. That is how Einstein arrived at his 1911 prediction. There is no sign of Newton in this, and yet the result is identical.

Masers and Lasers

A good clock must be stable over long time intervals. There are related devices that can achieve much greater stability over short time intervals. One of these is the hydrogen *maser*, which can maintain a stability, over periods of several hours, that is equivalent to one second in 100 million years. The word *laser* is probably more familiar. The maser is the forebear of the laser. The name is an acronym: m(icrowave) a(mplification) by s(timulated) e(mis-

Hydrogen maser built by the Smithsonian Astrophysical Observatory

sion) of r(adiation). Quite a mouthful. In the laser, l(ight) has replaced m(icrowave).

Masers and lasers are devices that raise atoms to a higher energy state, from which they are stimulated by suitable radiation to transfer that excess energy to radiation of exactly the same frequency, thereby amplifying it and producing radiation of an exceedingly sharp frequency.

Like cesium, hydrogen has a hyperfine structure, which produces microwave radiation of frequency 1,420 MHz, wavelength 21.1 cm.* In the hydrogen maser, a beam of hydrogen atoms is selected magnetically in the upper energy state and then directed into an evacuated chamber with nonsticky walls, where the atoms bounce around without significant disturbance. They are exposed to radiation of just the right frequency, which builds up an electromagnetic oscillation that, for several hours, holds steady to within less than a millionth of a Hertz.

Measuring the Red Shift Near Earth

In June 1976, a Scout rocket of the National Aeronautics and Space Administration was fired from the Atlantic coast on a nonorbital flight. It went to an altitude of about 10,000 km, a little more than one and a half times the radius of the Earth, and splashed down in the Atlantic, about 600 km southeast of its launching point. This was an experiment of Harvard's Smithsonian Astrophysical Observatory. The rocket carried a hydrogen maser. The output of this maser was continuously radioed to ground and there compared with that of an identically constructed maser. The frequencies differed. But almost all of this difference came from the Doppler effect produced by the motion of the rocket. To eliminate that, the signal of the ground-based maser was radioed to the rocket, where a device retransmitted it back to the ground. That essentially doubled the Doppler effect; so half of that frequency shift was automatically subtracted from the frequency difference of the two masers. When the rocket was near its maximum height, the net shift was about six parts in 10^{10}, a little less than 1 Hz. A major portion of that should be the gravitational effect. Indeed, the expected effect was verified to seven parts in 10^5.

*Because hydrogen plays a fundamental role in the universe, it has been suggested that this frequency might be preferred for signaling by extraterrestrial intelligent beings. CQ! CQ! Is anyone out there?

Epilogue

After servicing the orbiting Space Telescope, the space shuttle orbiter Atlantis *rendezvoused with, and recovered, the deep-space probe* Einstein. *Following reentry,* Atlantis *glided to Earth and touched down on the runway at Edwards Air Force Base, in California.*

It is rumored that Ein was quietly retired from active service, and is now receiving TLC somewhere in the Santa Monica Mountains.

THE PROBLEM

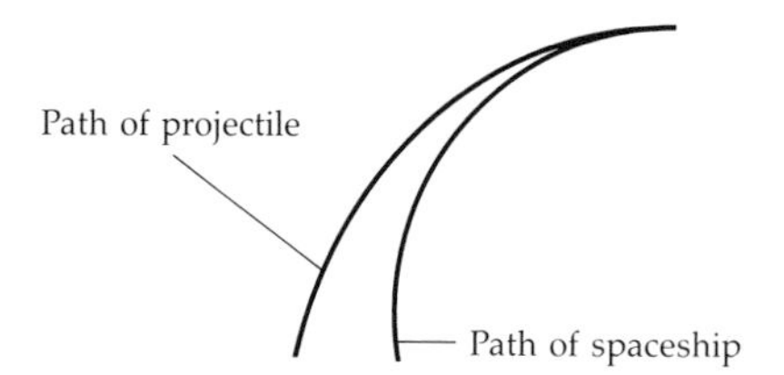

So ended "Mission Einstein." But the space age was far in the future when Albert Einstein, from 1907 until the final success at the end of 1915, labored to construct the *general theory* of *relativity*, which would give gravitation its rightful place in the world of space and time. The then-speculative Principle of Equivalence was a beginning, but only that. We have hinted at its failure in connection with the deflection of light by the Sun.

The problem that Einstein faced is brought out in the way that Ein finally realized that he was in orbit about the Moon. Recall that he fired a fast projectile. For a short time, projectile and spaceship fell together, the distance between them being still sufficiently small that they experienced the same gravitational force. But soon the fact that the gravitational field is not uniform, that it points to

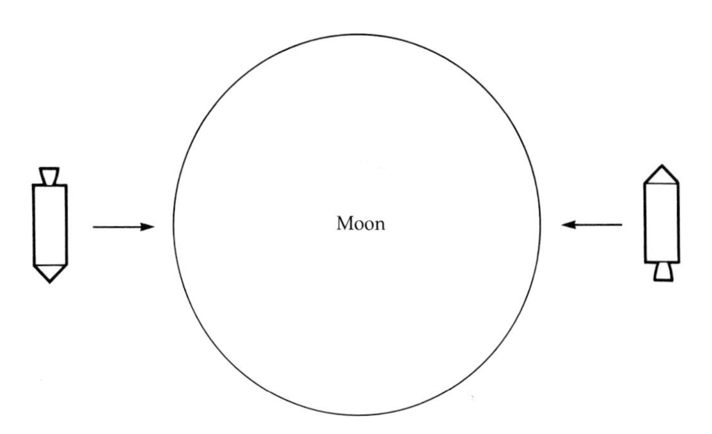

Different accelerated reference frames

the center of the Moon and weakens in strength with increasing distance, became significant. The projectile flew off, whereas the spaceship continued in its orbit. In short, the projectile disclosed what experiments within the ship could not, that a true gravitational field was at work.

There is the power, and the limitation, of the Principle of Equivalence. It asserts that, within a region of space and time that is small enough for the gravitational field to be uniform, all sign of that field disappears if we adopt a suitable accelerated frame of reference, one that is falling freely. Here the laws of special relativity apply: light travels at the speed *c*. But, relative to that reference frame, the laws of special relativity do *not* describe the phenomena near a significantly displaced point, the opposite side of the Moon, for example. There a *different* accelerated reference frame would be needed.

The problem is to unify all such limited, local descriptions into a connected whole. As described in the next chapter, the solution came only with the recognition that the geometry of space and time is not fixed independently of the matter in the universe but is determined by it. This is Einstein's crowning achievement.

NOTES

1. **H. Boorse and L. Motz, eds.:** *The World of the Atom* (Basic Books, 1966), p. 17.
2. **A. Einstein:** *Essays in Science* (Philosophical Library), p. 80.

5

GEOMETRY AND PHYSICS

GEOMETRY

The Greek traveler and historian Herodotus (484–425 B.C.) wrote:

> The king moreover (so they said) divided the country among all the Egyptians by giving each an equal square parcel of land, and made this his source of revenue, appointing the payment of a yearly tax. And any man who was robbed by the river of a part of his land would come to Sesostris and declare what had befallen him; then the king would send men to look into it and measure the space by which the land was diminished, so that thereafter it should pay the appointed tax in proportion to the loss. From this, to my thinking, the Greeks learned the art of measuring land . . . .[1]

Geometry (earth measurement) began in the mud of the Nile. In the hands of the Greeks—Euclid, the geometer of Alexandria (330–275 B.C.) became its eponym—it was divorced from reality, its foundations frozen in axioms. More than two thousand years would elapse before geometry was again brought down to earth. The liberator was Karl Friedrich Gauss, prince of mathematicians, physicist, engineer, astronomer, surveyor, who was not above surveying the Kingdom of Hanover (see Box 5.1). The new concept of geometry that Gauss was thereby led to invent received its full expression in the hands of Gauss's pupil, Bernhard Riemann. Many years later, the insight of Gauss and Riemann helped Einstein give mathematical form to his physical recognition that the geometry of space and time is *not* absolute and unalterable:, it is determined by the properties of matter, acting through the agency of gravitation.

Karl Friedrich Gauss

BOX 5·1 *Gauss*

Karl Friedrich Gauss was born April 30, 1777, in Braunschweig (Brunswick), the son of a sometime gardener, sometime bricklayer. The hand of his uncouth father lay heavy on the child. It was his mother's brother, a skilled weaver, who aroused Gauss's latent genius. The precocity of the child was shown before he was three. Hearing his father err in adding the sums for the weekly wages of his laborers, the little boy spoke up with the right answer. (It is said that Mozart had to age *four* years before composing his first minuet.) A preoccupation with numbers, and the ability to do complex calculations mentally, remained with Gauss throughout his life.

The phenomenon of the idiot savant—a calculating prodigy who is otherwise low in intelligence—is not unknown. Johann Dase, born in Hamburg in 1824, who came to Gauss's attention about 1840, was once asked to multiply 79,532,853 by 93,758,479; he gave the right answer in fifty-four seconds. He had no other talents. But *Gauss* was no idiot.

At the age of ten, after two years at a school run by a brutal teacher, Gauss entered a class in arithmetic. With the intention of keeping the class occupied for a whole hour, the schoolmaster gave an addition problem, in which a long series of numbers, each differing from the preceding one by the same amount, was to be added up. Almost before the other pupils began their laborious efforts, Gauss wrote one number on his slate and dropped it on the table before the teacher. Without previous knowledge, he had instantly sensed the simple rule for such sums: multiply the average of the first and last numbers by the total number of quantities to be added (see the footnote on page 176). The schoolmaster, properly amazed by this performance, changed his spots, and did his limited best to help Gauss. This ultimately led to Gauss's becoming a protégé of Duke Ferdinand of Brunswick at the age of fourteen. Gauss's education was now assured. Master of both languages and mathematics, the eighteen-year-old Gauss was unsure of which career to pursue when he entered the University of Göttingen. (Opened in 1737, the university had been founded by George II, King of Great Britain and Ireland, Elector of Hanover. From 1714 to 1837, a royal road ran between London and Hanover.)

A pentagon

Gauss made his decision at the age of nineteen with the solution of a two-thousand-year-old problem. The regular pentagon, a figure with five equal sides and five equal angles, can be constructed in the Greek manner—that is, by using only a straight edge and a compass. Can that also be done for other regular figures of seven, nine, eleven, . . . sides? (Like five, these integers are *not* the product of two smaller integers, each larger than one.) Gauss gave the general rule; the next possible example has seventeen

Gauss's observatory

sides; the Greeks missed it. This discovery is the first entry in a private notebook that remained inaccessible until 1898; in it were entered Gauss's anticipations of some of the major mathematical developments of the nineteenth century.

At the beginning of the nineteenth century, Gauss turned his attention to astronomy, which led in 1807 to his appointment as director of the Göttingen observatory. (His patron, Duke Ferdinand, had been mortally wounded a year earlier while serving with the inexperienced and underprovisioned Prussian forces at the battle of Jena-Auerstädt.) A new observatory, constructed under Gauss's direction, was completed in 1816; it still stands. Thereafter, and until his death in 1855, there were few nights that he did not spend under its roof.

On May 9, 1820, at Carlton House, London, the newly crowned George IV, King of the United Kingdom of Great Britain and Ireland, King of Hanover (the Holy Roman Empire, with its hereditary title of Elector, had ceased to exist in 1806), signed the order directing "Professor Gauss" to survey the Kingdom of Hanover in order to join up with the survey of the Duchy of Holstein, then part of Denmark. With the declaration that it was really all the same whether he measured the position of a star or of a church tower, Gauss characteristically threw himself completely into this earthy task. He devised new instruments for the purpose, and, of capital importance, he constructed the mathematical theory of curved surfaces that pointed the way for Riemann.

For further adventures of Gauss, see Box 6.1.

Euclid

Let us start with Euclid and the geometry of the plane, with its points, straight lines, circles, ellipses, and triangles. From Book I of the *Elements:*

Definition 4, I. A straight line is that which lies evenly between its extreme points.

It is quoted, not for its clarity, but its spirit. *We* would say that a straight line is the shortest *distance* between two points. We would go on to define a circle as the collection of all points at a given *distance* from a fixed point, an ellipse as the collection of all points at a given sum of *distances* from two fixed points. We also know that the angles in a triangle are determined by the *distances* between the three apexes of the triangle. In short, we take for granted the hard-won knowledge that Euclidean geometry is based on a certain concept of the *distance* between two points.

René Descartes (1596–1650)

Descartes

From Euclid, in Alexandria, to René Descartes, in Holland (see Box 5.2), is a span of almost two thousand years. That is how long it took before geometry progressed from being a set of special problems, each requiring individual ingenuity, to being a uniform and general method. The revelation had come eighteen years earlier, but June 8, 1637, was the date of the publication, in Leiden, of *A discourse on the Method of rightly conducting the Reason and seeking Truth in the Sciences: Further, the Dioptric, Meteors, and Geometry, essays in this Method*—mercifully known as, *The Method.*

BOX 5·2 ***Cogito Ergo Sum***

"I think, therefore I am."

That is René Descartes's celebration of intellect as the core of man's being. Mathematician and philosopher, he was born in 1596 at La Haye, near Tours, France, the son of a counselor to the local parliament. Of delicate health, the boy was self-educated until the age of eight, when he was sent to the Jesuit college at La Flèche in Anjou. There, he was granted the privilege of lying abed in the morning, and the practice of reflection in such horizontal circumstances was instilled in him for life. What he acquired in the eight years at La Flèche was *doubt.* At the age of sixteen, he left for the big city, Paris, first to engage in the social life centered on gaming and then to seclude himself in the pursuit of mathematics. His hideout discovered, he sought peace—in war. France at the time was no place for students or soldiers of honor, and so, in 1617, he set out for Holland (the Netherlands) to learn the trade of soldiering.

A truce then existed in the long struggle of the Netherlands for freedom from Spain and, after two years, Descartes sought activity in the army of the Duke of Bavaria. Snugly quartered on the banks of the Danube during the hard winter of 1619, he had time to meditate and was awarded mystical glimpses of his final philosophical system, in which his new concept of geometry played a major role. His further wanderings in and out of military service finally led him back to Holland in 1629. There he remained for twenty years, although, on the average, he changed his residence every ten months. Those years, which nearly coincide with the period of leadership by Frederick Henry, Prince of Orange, were perhaps the most brilliant of the Dutch Republic: the first colonists of Nieuw Amsterdam had landed in 1624; Rembrandt van Rijn's misnamed *Night Watch* dates from 1642. By 1633 Descartes was ready to publish *The World,* in which the Copernican system was accepted as a matter of course. And then word reached him of Galileo's fate. The book was put aside. "I desire to live in peace," he wrote.

The publication of *The Method* in 1637 (its essay titled "Meteors" gave the first scientific explanation of the rainbow), of *Meditations* in 1641, and of *Principles* in 1644 established the Cartesian movement.*

The growing fame of Descartes brought him, unhappily, to the attention of the young Queen Christina of Sweden. Nothing would do but he must create for her an academy of science. And, as it turned out, she also insisted on an hour of philosophical instruction each morning at the hour of five—in an unheated library! Poor Descartes hadn't a chance; he lasted only five months in the chill northern climate, failing by a month to reach the age of fifty-four. Christina's days on the throne were also numbered—she abdicated four years later.

*Cartesianism was elaborated and transformed by the Dutch philosopher Benedict de Spinoza (1632–1677). Einstein, who often made references to God, once explained that he had in mind the God of Spinoza: Nature.

What is the method? Suppose that you are in New York City, Borough of Manhattan, at the intersection of Fifth Avenue and Forty-second Street, in front of the public library. To meet someone at Seventh Avenue and Forty-fifth Street, you go two blocks west and three blocks north, in either order (ignore Broadway!). If you continue your walk, you can record your route by noting the avenue and street numbers of each intersection, which account can be made more precise by using building numbers (as in 111 Fifth Avenue). In short a *pair* of numbers can specify any position on the essentially planar surface of midtown Manhattan,* and the progression of that pair of numbers describes whatever path you may choose, without regard to whether Euclid mentioned that particular geometrical figure.

Midtown Manhattan

The library is just below the center of the picture.

*To avoid confusing this point, do not go east of Fifth Avenue and, as any New Yorker would do, speak of Sixth Avenue.

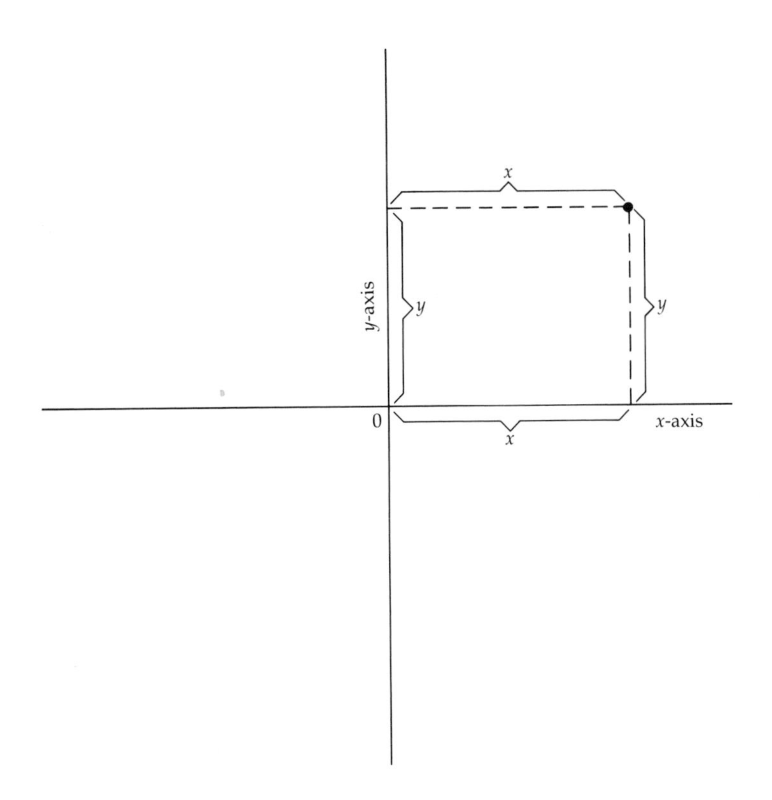

It may be hard now to realize that *someone* had to invent this method, or, at least, recognize its significance for geometry. There is a story that Descartes was lying in bed (this at least has the ring of truth—it was his favorite position when he wished to think) and watching a fly on the wing in the corner of the room. It suddenly occurred to him that the position of the fly at any instant was given by the three distances to the three planes formed by the intersecting walls and ceiling.

We turn to the simpler situation of a plane surface in order to present this inspiration in a slightly more useful form. Draw two intersecting straight lines, one horizontal and one vertical. This pair of perpendicular axes forms a reference frame. Any point on the surface is located with respect to this frame by two numbers, referred to generally as x and y. The first number, x, is the perpendicular distance to the vertical axis and is endowed with a plus or minus sign, according as the point stands to the right or the left of

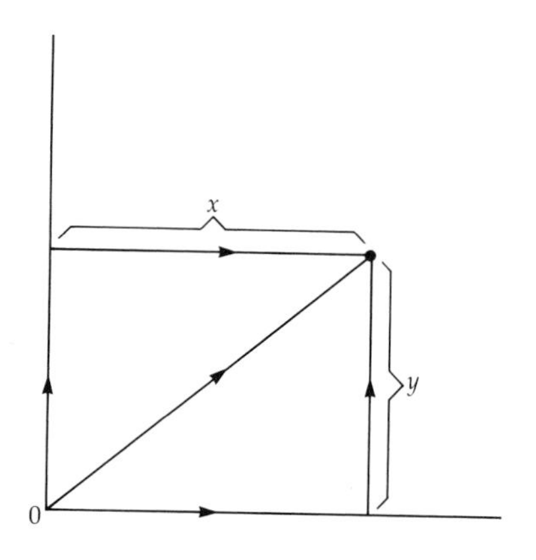

that axis. The second number, y, is the perpendicular distance to the horizontal axis; it, too, is endowed with a plus or minus sign, according as the point stands above or below that axis. The use of letters such as x and y follows Descartes, after whom they are called the *Cartesian coordinates* of the point. The point with coordinates $x = 0$, $y = 0$, the intersection of the two axes, is the *origin* of the reference frame.

Now consider some point with coordinates x and y. Only as a matter of convenience, take both coordinates to be positive numbers, which puts the point somewhere in the upper-right quadrant of the plane. You can reach that point from the origin by moving in a straight line to the point. You can also reach the point, just as in the walk from Fifth Avenue and Forty-second Street, by moving horizontally a distance x and vertically a distance y, in either order. The straight line to the point, along with the two perpendicular lines of lengths x and y, form a right-angle triangle. Then the Pythagorean theorem tells us that the square of the distance of the point from the origin is the sum of the squares of length x and length y, $x^2 + y^2$. Moreover, the square of the distance between *any* two points in the plane, labeled 1 and 2, is the sum of the squares of the *differences* in the values of x and y:

$$(x_1 - x_2)^2 + (y_1 - y_2)^2.$$

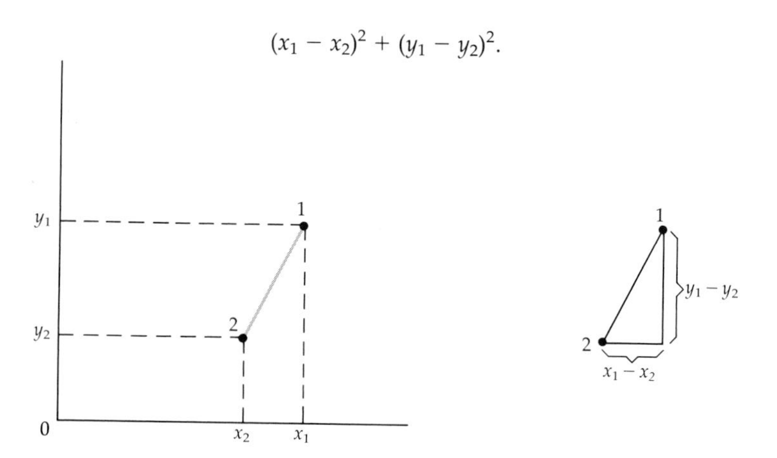

Perhaps the reason for the long historical delay in introducing the Cartesian reference frame is that it would have brought in something that seemed to be extraneous. But, in fact, absolute position in space is meaningless; all position is *relative* to other bodies. A frame of reference is an idealization of such bodies. The origin and the orientation of a reference frame can be chosen freely (from

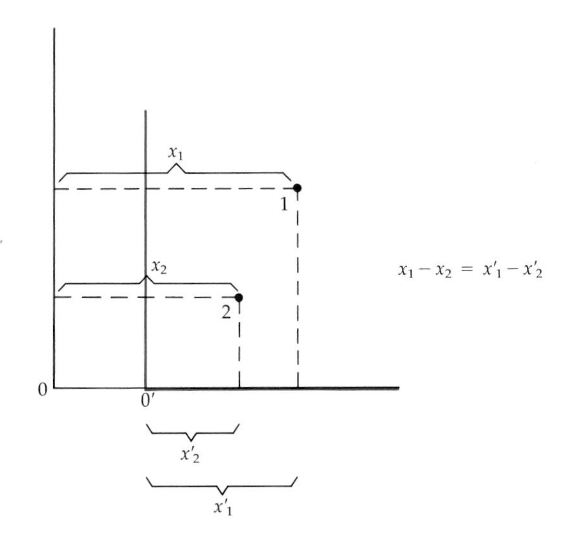

the beginning we have accepted the freedom of *motion* of reference frames). In the plane—the *two-dimensional* space of the coordinates x and y—displacing the origin along the x-axis, for example, changes the value of the x-coordinate for every point, by a common constant, but has no effect on coordinate *differences*, which describe the *relative* positions of points. Rotating the reference frame about its origin changes the values of both the x-coordinate and the y-coordinate for every point but has no effect on the value of $(x_1 - x_2)^2 + (y_1 - y_2)^2$, which gives the square of the distance *between* points.

Any path, or curve, can be described by giving the pairs of numbers, the coordinates x and y, that identify each point on that

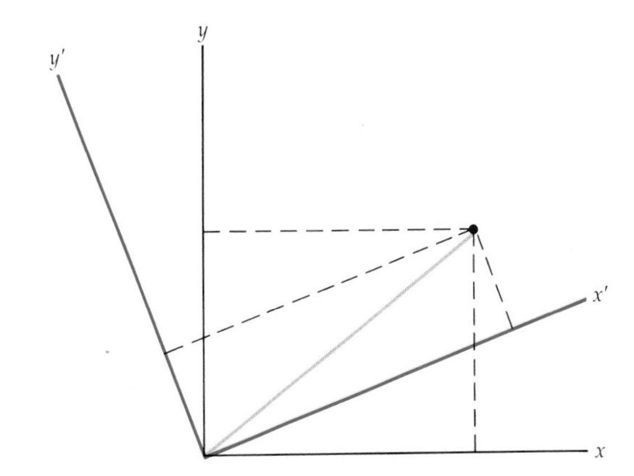

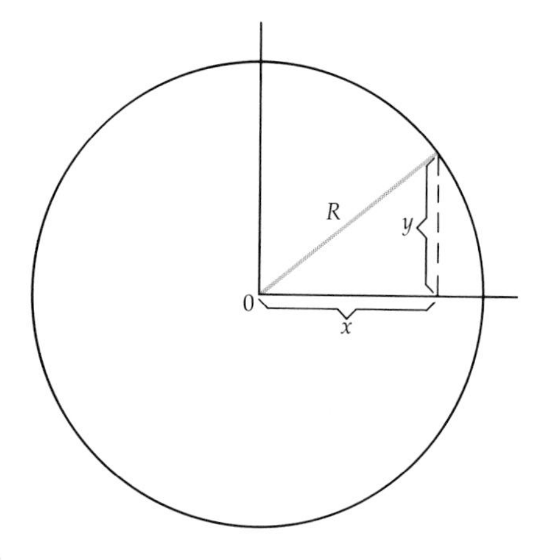

curve. A simple example is a circle of radius R. If we put the origin of a reference frame at the center of the circle, the definition of the circle as the collection of all points at distance R from the center becomes a relation between x and y:

$$x^2 + y^2 = R^2.$$

Any x,y pair that satisfies this relation gives a point on the circle; any x,y pair that does not satisfy it gives a point off the circle.

Another illustration is a straight line. Thus, $x = d$ (a positive constant) describes the collection of all points that are situated to the right of the y-axis, at the constant distance d; this is a straight line parallel to the y-axis. Then consider $x = y$, which describes the collection of points that, in the upper-right quadrant or the lower-left quadrant, are equidistant from the x- and the y-axes. This is a straight line that passes through the origin, inclined at a 45-degree angle to those axes. The general straight line is characterized by $x = ay + b$, in which a and b are freely chosen constants. (With the choices $a = 0$, $b = d$ and $a = 1$, $b = 0$, we regain the two special examples.) For any two points on a straight line, the difference in x values is proportional to the difference in y values; that constant of proportionality is 0 and 1, respectively, for the special examples.

It is a short step from the two-dimensional plane to the *three-dimensional* space that we inhabit (and in which Descartes and the fly were located). A three-dimensional reference frame has three mutually perpendicular axes intersecting at a point, the origin of the reference frame. The perpendicular distances from a given point to each of the three planes formed by various pairs of axes supply values of the coordinates x, y, z; again, these coordinates carry plus or minus signs to distinguish the two sides of each plane. Now, a twofold application of the Pythagorean theorem shows that $x^2 + y^2 + z^2$ is the square of the distance of the point from the origin. More generally, the square of the distance between *any* pair of points, again labeled 1 and 2, is

$$(x_1 - x_2)^2 + (y_1 - y_2)^2 + (z_1 - z_2)^2.$$

Displacements of the reference-frame origin do not change the coordinate differences that give relative positions; rotations of the reference frame do not change relative distances.

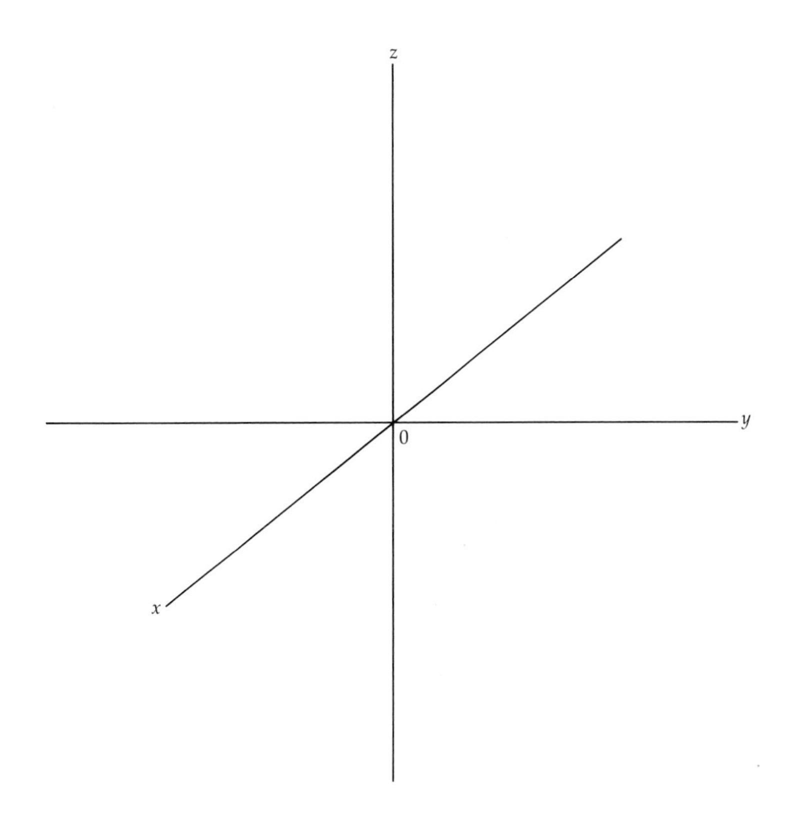

A three-dimensional reference frame

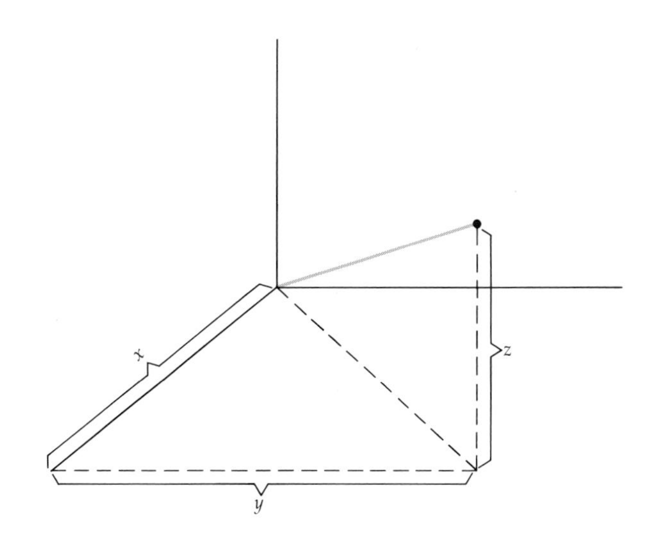

Coordinates of a point in a three-dimensional reference frame

The Sphere

A sphere is the collection of all points in three dimensions at a given distance from a fixed point. Let that distance, the radius of the sphere, be called R. If the fixed point, the center of the sphere, is chosen as the origin of the coordinate frame, the points of the sphere are such that

$$x^2 + y^2 + z^2 = R^2.$$

This is the description of the surface of the sphere as an object in three-dimensional space. But the surface itself is *not* three-dimensional. *Two* coordinates locate any point on the surface. The use of latitude and longitude is familiar. The circles of latitude begin at the equator, which is labeled zero degrees (0°), and progress with decreasing circumference to the North Pole (90° N) and to the South Pole (90° S). Intersecting those circles at right angles are the circles of longitude, which pass through the poles. These circles begin at the Greenwich meridian (0°) and proceed through 180 degrees to the west and the east.

The sphere is a two-dimensional surface imbedded in three-dimensional Euclidean space. To a sailor navigating the seas of the real Earth, or to a surveyor measuring its land, all interest is focused on what is found *on* the surface, as observed from that surface. Is the geometry of the Earth's surface—or, more simply, the geometry of a sphere—Euclidean geometry?

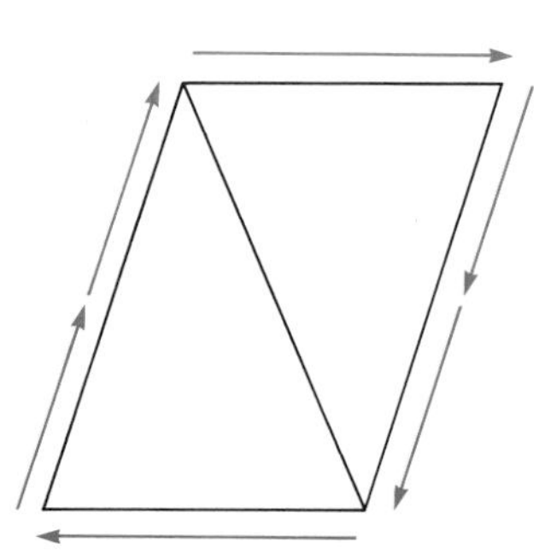

How shall we recognize Euclidean geometry? Here are some signs. Draw a straight line of a certain length. At the two ends of that line draw two other straight lines, of a common length, that make the same angle with the initial line. Then the two ends of those lines can be connected by a straight line of the *same length* as the first line. Start somewhere on one of the lines of the closed four-line figure. Follow these lines through two successive angles; you are now moving in the opposite direction, and have swung through half the angle of a full circle, or 180 degrees. Completing the circuit to your starting point doubles that; so the four angles of the figure add up to 360 degrees, a *full circle.* Now divide the figure in half by a diagonal straight line that connects opposite corners. This produces two triangles that can be rotated and perfectly matched to each other; they are *identical* triangles, which can be drawn to any prescribed scale. Therefore the sum of the angles of *any* triangle is *half* that of a full *circle,* or 180 degrees. As for circles,

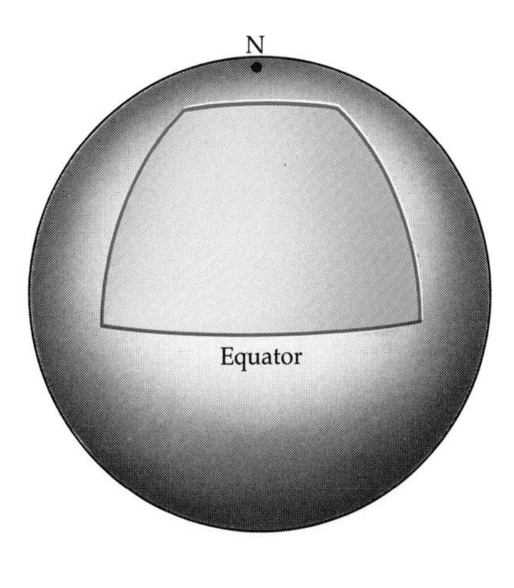

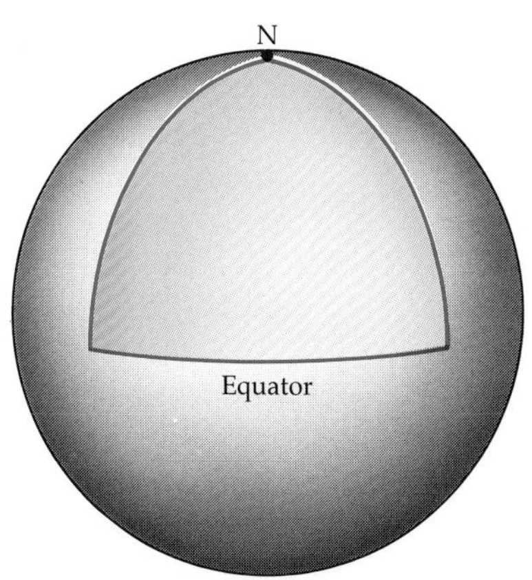

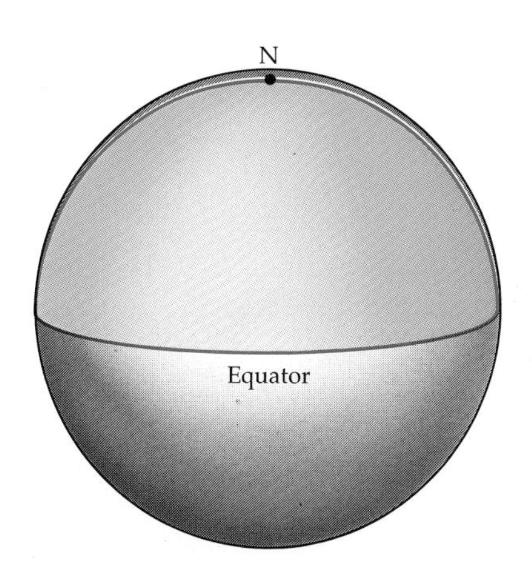

the ratio of the circumference to the radius of *any* circle is 2π, the angle of a *full circle* measured in radians (see Chapter 4).

The long-held belief that the Earth is flat, a plane surface, is testimony of a kind that a *very small* area of a sphere (or indeed of any surface) is effectively Euclidean in its geometry. If it is accepted that the Earth is essentially a sphere, is the geometry of its *whole* surface Euclidean or not?

Let us draw on the surface of a sphere a *straight line,* which we continue to define as the shortest distance between two points. For example, a string that connects two points and is drawn as tightly as possible lies on a straight line between them. If the points are on the equator, the string will follow the line of the equator, which is a great circle. (A great circle is such that a slice through it divides the sphere in halves.) There is nothing special about the equator; any straight line on a sphere is a part of a great circle. (A more general term for a line of shortest distance is *geodesic,* "earth divider.")

Start with that straight line along the equator and, at each end, draw perpendicular straight lines of equal length, to the north. These lines are parts of the great circles perpendicular to the equator, which are lines of longitude. Now, connect their two ends by a straight line. Is the length of that line equal to the length of the initial line on the equator? The closer the ends of the lines are to the North Pole, the clearer it is that the test *fails.* The lines of longitude converge as we proceed north; all of them intersect at the pole. The length of that fourth line diminishes and approaches zero as its ends close in on the pole.

Suppose that we let those lines from the equator run all the way to the North Pole? Then we get a triangle. Let us check the sum of its three angles against 180 degrees. At the equator, where the lines to the North Pole are at right angles to the equator, we already have 180 degrees; thus, the sum of the three angles *exceeds* 180 degrees by the value of the angle at the polar apex. As we increase this polar angle, the area of the triangle increases proportionally. When the apex angle is 180 degrees, or π radians, the triangle is one-half of the northern hemisphere, one-quarter of the total surface area of $4\pi R^2$, which is πR^2. This shows that the *excess* of the sum of the angles over π radians is the *area* of the triangle divided by R^2. The same relation holds for any triangle on a sphere.

How about circles? Draw one centered on the North Pole, producing a fixed radius by a taut string extended to the equator. The

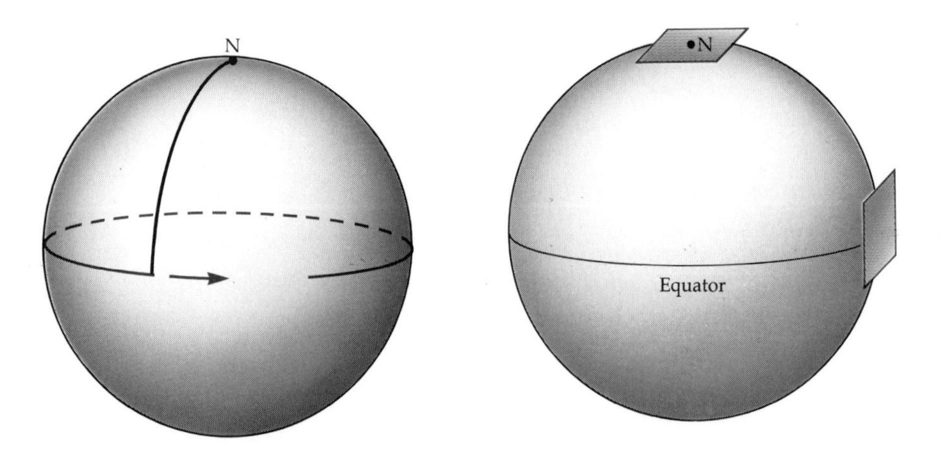

Drawing a circle on a sphere (left); Cartesian reference frames at the equator and the North pole (right)

circumference of this circle is the circumference of the sphere of radius R; and the radius of the circle, the length of the straight line from pole to equator, is one-quarter of that circumference. The ratio of circumference to radius is not $2\pi = 6.28\ldots$, but 4. What happens if we run the string from the North Pole to the South Pole? Then the circle is just the point at the South Pole, and the ratio of circumference to radius is *zero*!

The sphere fails the tests. The geometry of the curved spherical surface is, to use a term introduced by Gauss, *non-Euclidean*. The fact that the geometry *is* Euclidean over any very small part of the surface is not in conflict with this. The Cartesian reference frame that applies in one small Euclidean region—say, at the equator—is *not* the same frame that applies in a significantly different region—say, at the North Pole.

Non-Euclidean Geometry

The foundations of Euclidean geometry underwent increasing scrutiny after the rebirth of learning in the Renaissance. The veracity of Euclid's *Elements* as a whole was not at issue; rather, it was the credibility of a particular axiom, the last of Book I. A paraphrase of the lengthy Greek original is this: Given a straight line and a point off that line, one and only one line can be drawn, through the point, that is parallel to the given line. To later geometers, this axiom was not self-evident, nor was it obvious that it could not be deduced from other, accepted axioms. Girolamo Saccheri showed, in a work published at Milan in 1733, that the axiom of parallels would follow if there were just one triangle that had the sum of its angles equal to 180 degrees. In attempting to demonstrate the necessary truth of this, he considered the non-Euclidean alternatives, which he then mistakenly believed that he had disproved.

Gauss began his lifelong preoccupation with the foundations of geometry at the age of twelve; four years later he had the first suspicion that Euclid was not the last word. In 1817, when he was forty, he wrote:

> "I became more and more convinced that the necessity of our geometry [Euclidean] cannot be demonstrated, at least neither by, nor for, the human intellect. In some future life, perhaps, we may have other ideas about the nature of space which, at present, are inaccessible to us. Until then we must consider geometry as of equal rank, not with arithmetic, which is purely logical, but with mechanics, which is empirical.[2]

Gauss is saying that the true geometry of space must be found by experiment. A few years later, he did such an experiment. The measurement of the angles of a triangle over distances of 70, 87, and 107 kilometers, disclosed no discrepancy with Euclid.

We know that Gauss had devised an example of non-Euclidean geometry, but, as he wrote in 1829, he had not published it for fear of "the clamor of the Böotians," a reference to the people of a region in ancient Greece who were renowned for their obtuseness. The first publication of a non-Euclidean geometry, also in 1829, came from remote Russia. The author was Nikolai Ivanovitch Lobatchevksy, professor and rector at the then relatively new University of Kazan.* First announced in 1826, this work describes a geometry in which there are an unlimited number of lines, through any point, that do not intersect a given line. In this geometry, the sum of the angles of a triangle is *less* than 180 degrees; that was also true of the geometry invented by Gauss.

Non-Euclidean geometry was in the air. János Bolyai, a Hungarian army officer, had reached similar conclusions in 1823 and

*The town is now the capital of the Tatar Autonomous Soviet Socialist Republic; it is located on the Volga River, 700 km east of Moscow.

published them in 1831 as an appendix to a work of his father, Wolfgang Bolyai. Remarkably enough, Wolfgang was an early friend of Gauss, and yet there is no overt evidence that the reticent Gauss had ever mentioned to Bolyai his heretical attitude toward Euclid. Still, Gauss never publicly acknowledged the contribution of the younger Bolyai, whereas, on becoming aware of Lobatchevsky's work, about 1840, he intervened to ensure that Lobatchevsky was elected a member of the Göttingen Academy in 1842.

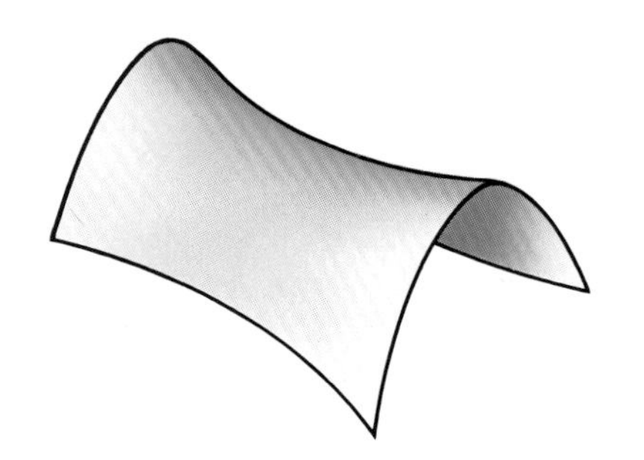

It is interesting that all three discoverers of non-Euclidean geometry, Gauss, Lobatchevsky, and Bolyai, came upon a geometry in which the sum of the angles of a triangle is *less* than 180 degrees, in contrast with the apparently more obvious example of the sphere,* where the sum of the angles exceeds 180 degrees. The kind of space in which the sum of the angles in a triangle falls short of its Euclidean value is found on the surface of a saddle, where the sense of curvature in one direction is opposite to that of the perpendicular direction (more will be said about this shortly.)

THE ROYAL ROAD

There is a tradition that, inscribed over the door of Plato's Academy outside of Athens, was "Let no man ignorant of geometry enter here." One who had no difficulty in crossing the threshold was Manaechmus, the discoverer (about 350 B.C.) of the ellipse, parabola, and hyperbola as sections of a cone (although the names come to us from a later Alexandrian, Apollonius of Perga). Almost two thousand years would pass before Kepler recognized the ellipse in the heavens.

It is said that Alexander the Great, conqueror of all the known world, asked Menaechmus for an easy path to the conquest of geometry. The reply was: "O King, for traveling over the country there are royal roads and roads for common citizens, but in geometry there is one road for all."[3] This democratic sentiment is often expressed in the shorter antiroyalist form, "There is no royal road

*Perhaps there was an unconscious aversion to a geometry in which there are *no* lines parallel to a given line. On a sphere, any two great circles intersect—twice—as illustrated by the great circles of longitude, all of which intersect at both poles.

Alexander at the Battle of Issus (Naples Museum)

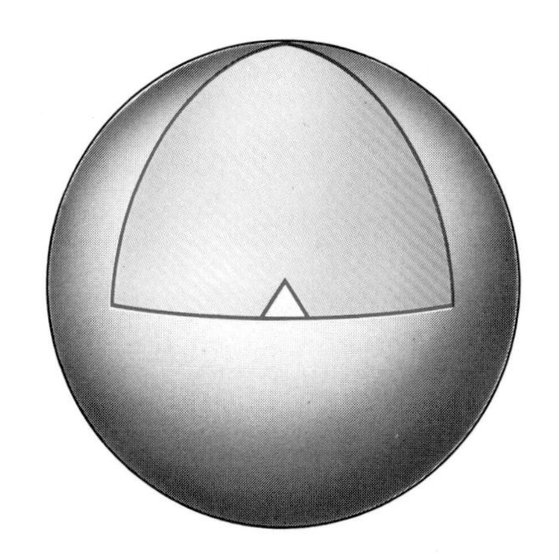

to geometry," and is sometimes placed later in history, with Euclid and Ptolemy I as the protagonists. However this may be, there *is* a royal road to geometry. Discovered by Karl Friedrich Gauss, its paving blocks are measuring sticks.

Over a sufficiently small area of any surface, the geometry is Euclidean. For a triangle drawn on a sphere, the amount by which the sum of the angles exceeds π is equal to the area of the triangle divided by the square of the radius of the sphere. When all the dimensions of the triangle are very small compared with that radius, this excess becomes negligible, and geometry over such limited distances is Euclidean. Then we can draw Cartesian axes, and we can compute squared distances as the sum of the squares of coordinate differences. But, as noted earlier, such a reference frame, laid down in one small area of the sphere, has no meaning for a small area in a significantly different region of the sphere.

Coordinates that do have meaning over the whole sphere, such as latitude and longitude, are not Cartesian coordinates. They are called, appropriately, *Gaussian coordinates.* Cartesian coordinates have an immediate interpretation in terms of distance: increase x by one unit, and the point in question moves parallel to

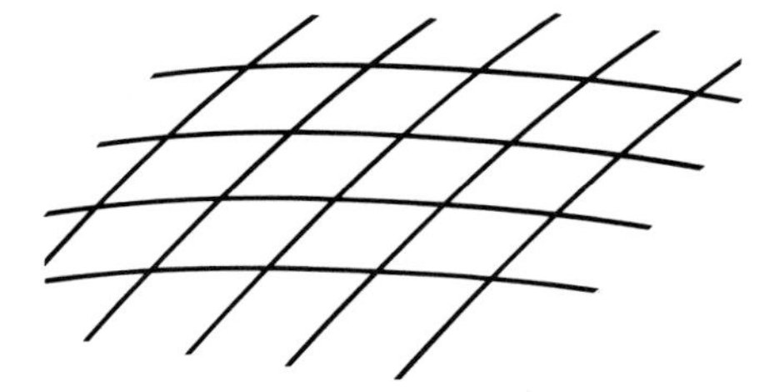

the x-axis by one unit of distance. It was Gauss who recognized that, in dealing with curved surfaces, we must give up this simple connection and, instead, use coordinates as *labels*. Draw two systems of curves in the surface, so that different curves of the same type do not intersect (except possibly at isolated points) and each curve of one type intersects all the curves of the other type. Then an arbitrary point is identified by the labels of the particular pair of curves that intersect there. Think of the Earth. One circle of latitude does not intersect any other circle of latitude. A circle of longitude does not intersect any other circle of longitude (except at the poles). Each circle of latitude intersects every circle of longitude, and conversely. Those intersections happen to be at right angles, but this is a special circumstance attributable to the symmetry of the (almost) spherical Earth. It is also a special circumstance that each change in latitude by one degree always represents the same distance on the spherical Earth, about 110 km. There is no suggestion yet that latitude and longitude are not Cartesian coordinates. The truth emerges when we look at the distance traversed in changing the *longitude* by one degree. At the equator, it is also about 110 km but, at the latitude of New York (41° N), it is only 83 km. At latitude 89° N, 110 km from the North Pole, one degree of longitude is 1.9 km.

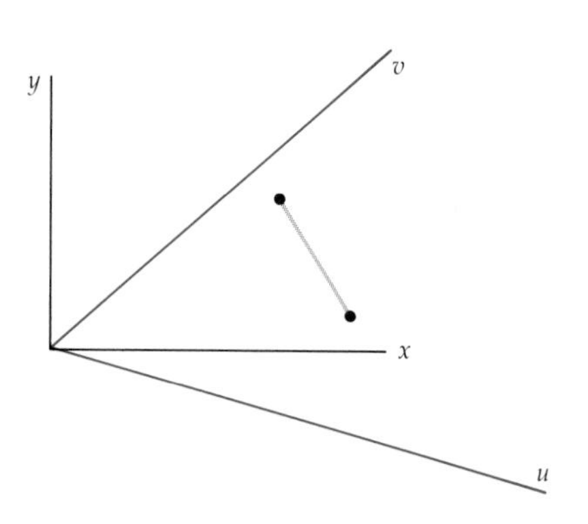

If Gaussian coordinates do not generally have a simple meaning in terms of distance, how do we compute distances? Think of a small area of a surface. The surface as a whole can be described by some kind of Gaussian coordinates, which we call u and v. Within the small area we can also adopt a pair of Cartesian coordinates, x and y. Now let us move from one point to another within the small area. This displacement can be expressed either by a small change of x and y or by a small change of u and v. The small changes of x and y are related by certain multiplicative factors to the small changes of u and v. Those factors are quite constant within the small area but generally vary as one progresses over the whole surface. The squared length of the straight line that connects the initial point to the displaced point is the sum of the squares of the changes in x and y. And that sum in turn can be written as the sum of *three* contributions: a multiple of the square of the small change in u; a multiple of twice the product of the small changes in u and v; a multiple of the square of the small change in v. The three multiplicative coefficients (Gauss designated them by E, F, G) also are constant within the small area but generally change as one moves over the whole surface.

These three* coefficients defined over the surface completely characterize the results of *measurements* made in the surface; accordingly, we refer to them collectively as the *metric* (of the surface). Thus, an angle, which is specified by the *lengths* of the sides of a *small* triangle, is determined by the metric at the location of the apex. The area of a *small* rectangular region, which is the product of its perpendicular *lengths,* is determined by the metric at that location. The area of any *large* section of the surface is found by adding such small bits of area. We can work out the finite distance between any two points, measured along a given connecting curve, by adding the lengths of the small line segments into which the curve can be divided. Then we can look for the *curves of shortest* length, the *geodesics* of the surface. And, armed with the geodesics—the straight lines of that surface—we can now construct *large* triangles and measure their angles to test whether the surface is Euclidean or non-Euclidean. In more intuitive language, we can now ask whether the surface is flat or *curved.*

Curvature

Recall that, on the curved surface of a sphere with radius R, the sum of the angles of a geodesic triangle exceeds π radians by the area of the triangle, multiplied by $1/R^2$. Now $1/R^2$ is clearly a measure of the *curvature* of the sphere; the smaller the radius, the more curved the sphere, and the larger the value of $1/R^2$. The *Euclidean* limit, in which the sum of the angles of a given triangle equals π, is the limit in which $1/R^2$ becomes *zero.* It is approached as R becomes very large. Why is this curvature the *square* of $1/R$ rather than just $1/R$? Because the surface has *two* dimensions. If you cut into the sphere perpendicularly at some point, you get an arc of a circle, the curvature of which is naturally measured by $1/R$. But you can make a second cut at that point, at right angles to the first one, and

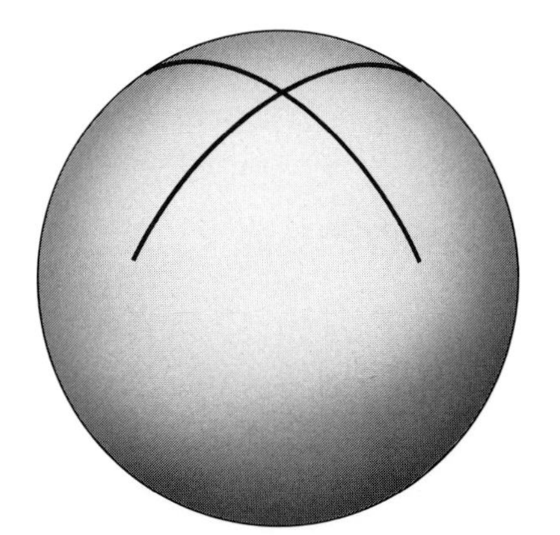

*Are we sure that all three coefficients are needed? Perhaps there are some general relations among them, so that a smaller number would suffice? Not if we wish to have free choice of the two systems of labeled curves that supply Gaussian coordinates. For a given v, a *small* change in u moves the point along a *straight* line. We are free to assign a number to the length of that line. Similarly, the length produced by a *small* change in v, for a given u, can be freely assigned a number. How about the angle between those two straight lines? That also can be freely assigned. This *three*-fold freedom of choice requires that *three* coefficients be specified, within each small area.

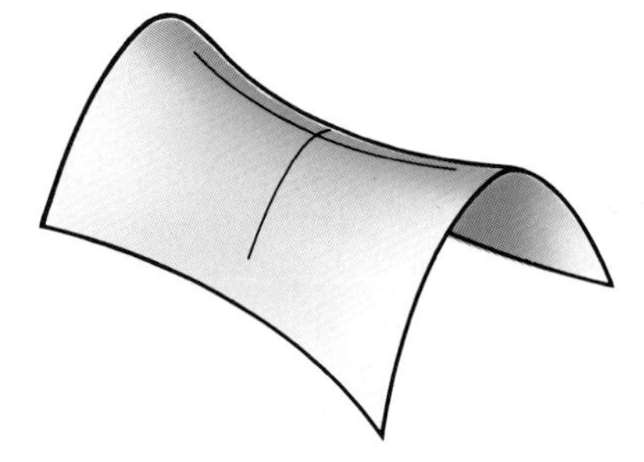

produce another circular arc of curvature $1/R$. It is the product of these one-dimensional curvatures that forms $1/R^2$.

The correctness of this interpretation becomes clear when we consider a saddlelike surface. Cut it transversely to the axis of the saddle, and we get a circular arc, as with the sphere. But a longitudinal cut, along the axis, gives a curve with the *opposite* sense of curvature. That change in sense is represented by a *minus* sign. And then the two-dimensional saddle curvature, given by the product of the curvatures for the two perpendicular directions, is *negative.* Now recall that the sum of the angles of a triangle differs from π by the area of the triangle multiplied by the *curvature.* Then, if the curvature is *negative,* as it is for the saddle, the sum of the triangle angles should be *less* than π radians, which it is.

Here, then, is a sensible definition of curvature. And it has a remarkable property: the "egregious theorem," published by Gauss in 1827, which states that the curvature so defined is completely determined by the *metric.* In other words, the Gaussian curvature is an intrinsic characteristic of the surface, quite independent of how the surface appears in three dimensions. That property is evident from the meaning of the curvature at a point in terms of the angles and the area of a very small triangle, which are quantities measurable *in* the surface. As an intrinsic characteristic of the surface, the Gaussian curvature at a point cannot depend on the particular choice of Gaussian coordinate system used to label that point.

Here is an example that shows how we can be misled by intuition or, perhaps we should say, the use of words in common currency without careful regard to precise definitions. A sheet of paper lies on a desk. It is a flat surface; its geometry is Euclidean. The angles of a triangle drawn on it will add up to 180 degrees. Pick

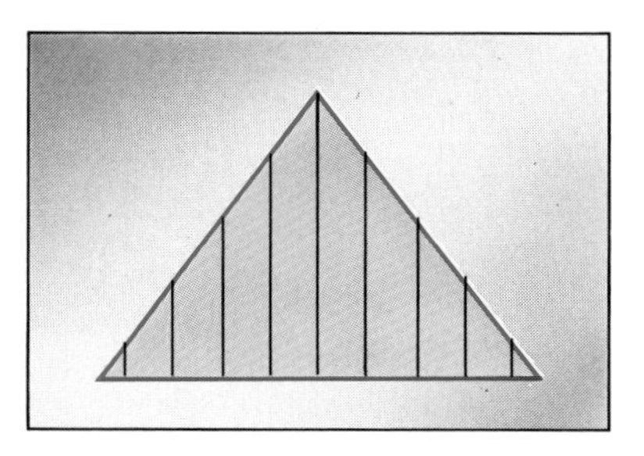

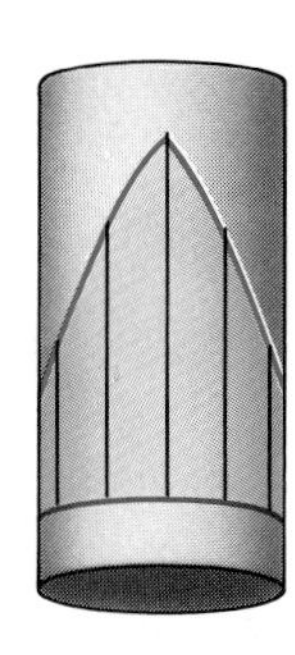

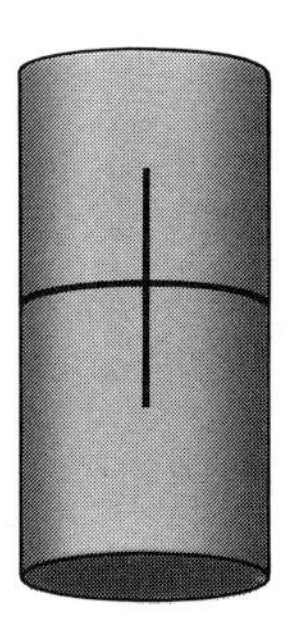

up the sheet and roll it into a nonoverlapping circular cylinder. In the usual sense of the word "curved," the surface of the cylinder is curved. You might well think that the geometry of the surface is non-Euclidean. Not true; the angles of the triangle in the cylindrical surface still add up to 180 degrees.

What counts here is not the feeling of curvature, but the precise value of *Gauss's* curvature. If we cut the cylinder transversely, we get an arc of a circle; its curvature is the inverse of the radius of the circle. Fine, but how about the cut along the axis of the cylinder? *That* is a straight line; its curvature is *zero*. Then Gauss's curvature, the product of the two one-dimensional curvatures, is *zero*. The geometry of the two-dimensional cylindrical surface is Euclidean, no matter how it looks in three dimensions.*

Riemann

In 1853, toward the end of Gauss's life, his pupil Bernhard Riemann, then twenty-seven, was required to address the Göttingen faculty to secure a position as lecturer, for which his recompense would come only from the fees paid by the students who elected to attend his lectures. It was customary to submit three possible topics, with tradition dictating that the choice would be made between only the first two. Those he had fully prepared. His third topic, the foundations of geometry, he had not prepared. But Gauss, who had grappled with the foundations of geometry for more than sixty years, could not resist that third topic. This was too much for Riemann, then immersed in investigations on electricity, magnetism, light, and gravitation (alas, this does *not* number among his claims to fame) while acting as assistant to Wilhelm Weber's seminar. He fell ill. After Easter of 1854, he recovered and in seven weeks completed his lecture. Gauss, feeling weak, postponed the lecture date, and then, suddenly set it for 11:30 in the morning of Friday, June 10, 1854.

Georg Friedrich Bernhard Riemann (1826–1866)

The title of Riemann's epochal lecture was "On the hypotheses which lie at the foundations of geometry." What did Riemann accomplish here? To those who are geometrically inclined, two di-

*There is no denying that there are *differences* between a plane surface and a circular cylinder. If, on a plane surface, you move away from a point, along any straight line, you never return to the initial point. Moving on the surface of the cylinder, perpendicularly to its axis, eventually brings you back to the starting point.

mensions is a breeze, three dimensions routine, and four dimensions impossible. But, to those who think algebraically, two, three, or four dimensions are just particular examples of spaces with *any* number of dimensions. In this sense, Riemann was an algebraist. He extended Gauss's intrinsic geometry of two dimensions, requiring two coordinates, to spaces of any number n of dimensions and coordinates: $n = 2,3,4, \ldots$. The metric of these spaces, which is specified by three quantities for $n = 2$, requires six for $n = 3$, ten for $n = 4$, and so on. To see this for $n = 3$, add a third coordinate to a two-dimensional space. The square of the distance between nearby points now contains the square of the change in the third coordinate, with a multiplicative coefficient, along with the products of the change in the third coordinate by the changes of each of the first two coordinates. That is three additions to the initial three coefficients, giving six quantities for $n = 3$. Proceeding to $n = 4$, we must similarly add one plus *three,* or four coefficients, to the six of $n = 3$, giving ten in all. And so it goes.*

It is amusing that here we have come back to the mystical beginnings of arithmetic in the hands of Pythagoras and his followers. At a time when counting was done with the use of pebbles (this survives in our language through Latin: *calculus,* a pebble),

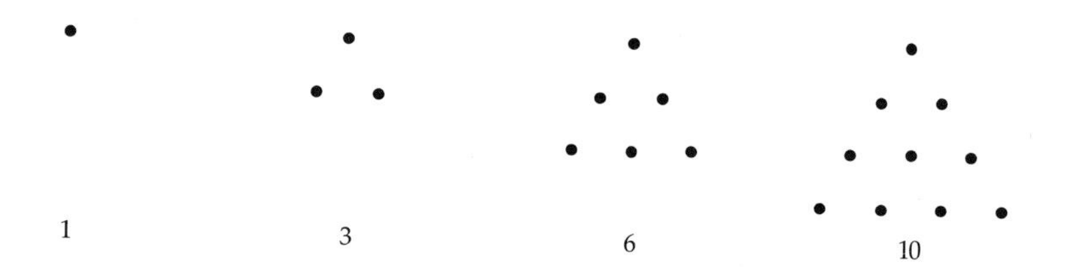

*The number of quantities that are needed to specify the metric in n dimensions is $1 + 2 + 3 + \cdots + n$, or (see Box 5.1)

$$\frac{1}{2}n(n + 1).$$

The structure of this expression is such that on adding $n + 1$ to it we get back the same form, with n replaced by $n + 1$:

$$\frac{1}{2}n(n + 1) + n + 1 = \frac{1}{2}(n + 1)(n + 2).$$

Here, for any n, is just what we did in going from two dimensions to three, and from three dimensions to four.

special significance was attributed to the numbers that count the pebbles in triangular arrays. The successive triangular numbers 1,3,6,10 . . . are produced as the equal number of pebbles on each side is increased by one. Of particular interest to the Pythagoreans was the fourth triangular number, the *tetractys:* $10 = 1 + 2 + 3 + 4$. It represented the universe.*

Riemannian Geometry

Given the metric of an n-dimensional space, what sort of things do you do in n-dimensional geometry, Reimannian geometry? Just what you do in two-dimensional Gaussian geometry: work out angles, areas (which might better be called volumes† now), geodesics. And then we come to *curvature.* For $n = 2$, Gauss had discovered the intrinsic curvature at each point. Riemann built on that discovery. From a point, draw, in some direction, a very short length of geodesic. Then pick a perpendicular direction (one at a right angle) and draw another short length of geodesic from the point. The two perpendicular lines define a two-dimensional surface. It has a Gaussian curvature. If the space itself is two-dimensional, we stop there. But, if it has three dimensions, $n = 3$, we can erect a third perpendicular geodesic line at that point. Then, in addition to the first two-dimensional surface defined by geodesics, we have two more, each with its Gaussian curvature. So, for spaces of dimensionality $n = 2$ and 3, the number of Gaussian curvatures is one and three, respectively. One and three: yes, we meet the triangular numbers again.

For dimensionality n, the number of different Gaussian curvatures, based on some choice of n mutually perpendicular geodesics at a point, is $\frac{1}{2}n(n - 1) = 1, 3, 6, \ldots,$ for $n = 2, 3, 4, \ldots$. Although these individual curvatures depend on the geodesics that

*In Pythagorean cosmology, which later influenced Copernicus, a central fire (unseen from Earth) was surrounded by *ten* moving bodies. The nearest to the center (also unseen) was the Antichthon, or counter-Earth, next came the Earth, then the Moon, the Sun, the five known planets, and last, the fixed stars. All shone by the reflected light of the central fire. Moving at speeds that increased with their distance form the center, the ten bodies gave out successively higher musical tones, which combined harmoniously into the (unheard) cosmic music of the spheres. It was a beginning.

†From the *area* of a small rectangular region, defined as the product of its two perpendicular lengths, we go to the *volume* of an analogous small region: the product of its n perpendicular lengths. Such a measure in terms of length does not depend on the particular choice of coordinates.

define the various two-dimensional surfaces, the sum of all $\frac{1}{2}n(n-1)$ of them does *not*. This sum supplies a kind of mean curvature at the point, one that depends only on the metric. If the space is Euclidean, or flat, which requires that all the Gaussian curvatures vanish everywhere, the mean curvature is zero. But it does not work the other way; the mean curvature can vanish everywhere without the space being flat. And the mean curvature at a point does *not* depend on how the system of n-dimensional coordinates used to label the point is chosen.

PHYSICS

In concluding his lecture, Riemann said:

> The question of the validity of the hypotheses of geometry in the infinitesimally small is bound up with the question of the basis of the metrical relations of space. . . . we must seek the basis of its metrical relations outside it, in binding forces which act upon it.
>
> A decisive answer to these questions can be obtained only by starting from the conception of phenomena which has hitherto been justified by experience, and which Newton assumed as a foundation, and then making in this conception the successive changes required by facts which it cannot explain; researches of this kind, which commence with general notions, can be useful in preventing the work from being hampered by too narrow views, and in keeping progress in the knowledge of the interdependence of things from being checked by traditional prejudices.
>
> This leads up into the domain of another science, that of physics, into which the character and purpose of the present discussion does not allow us to go today.[4]

That was spoken in 1854. Sixty years would elapse before the nature of the "binding forces" that determine the metric of geometry was understood. They are the forces of *gravitation*.

Riemann died at the infuriatingly early age of thirty-nine. Had he been granted another twenty years, until July 20, 1886, almost a year before the Michelson-Morley experiment, would he have produced Einstein's theory of gravitation? The reluctant but very probable answer is: no. Reimann was thinking of space—three-dimensional space. Lacking was the whole physical development, culminating in the special theory of relativity, that linked *time* with space. Three-dimensional geometry must give way to a four-dimensional geometry in the relativistic world. What, then, is the space-time geometry of special relativity?

Space-Time and Geometry

Geometry has been based on the metrical concept of distance, a relation between two points that is independent of the frame of reference. Do we already know something that, unlike relative temporal and spatial intervals, has this character of absoluteness? Yes. It was identified back in Chapter 2, although now it is more convenient to change its algebraic sign; that quantity is $L^2 - (cT)^2$, in which L and T are the spatial and temporal intervals between two events, labeled 1 and 2. Let each event be assigned spatial Cartesian coordinates, x, y, z, and time t. Then the absolute quantity is

$$(x_1 - x_2)^2 + (y_1 - y_2)^2 + (z_1 - z_2)^2 - (ct_1 - ct_2)^2.$$

Had the algebraic sign of the fourth term been positive, this would be the squared distance of four-dimensional *Euclidean* geometry. The geometry of special relativity is four-dimensional Euclidean geometry, with a *difference*. The causal circumstances in which this absolute quantity could be positive, zero, or negative were presented in Chapter 3. *Negative* squared distance may be a geometrical nightmare, but it poses no serious algebraic problems. Although this space-time geometry is not exactly Euclidean, the four dimensional space is certainly *flat*. The ten coefficients that specify the metric are fixed numbers; variously, 0, +1, and −1. There is *no* change of this metric in moving from one point to another, which is the hallmark of *zero* curvature.

The Principle of Equivalence and Geometry

The Soviet space station Salyut 8 is in orbit about Earth. In a different, higher orbit, the U.S. space-shuttle orbiter Columbia is carrying in its payload bay the European Space Agency's Spacelab. Within one section of it, four specialists, working in shirt sleeves, are monitoring several experiments being performed in the other section, under direct exposure to the space environment. To the occupants of each spaceship, there is no sign of Earth's gravity or of the acceleration of free fall. The two effects precisely cancel. The equivalence (indistinguishability) of a constant acceleration and a uniform gravitational field was raised to a principle by Einstein. It

Soviet space station

asserts that, within sufficiently small regions of space and time, described relative to suitably accelerated frames of reference (here, the freely falling spaceships), all physical effects of a gravitational field are removed, and the laws of special relativity apply. What happens inside the spaceships demonstrates the principle: give an object—say, a handy wrench—a push, and it moves in a straight line at constant speed (until it hits something). A beam of light also moves in a straight line, traveling at the characteristic constant speed c.

Special relativity applies within each spaceship, relative to its own freely falling frame of reference. But the two accelerated frames of reference are quite distinct; that is particularly apparent when the different orbital paths place the spaceships on opposite sides of the Earth. Not only that, the measure of time differs between the two spaceships; one tick of a clock does not have the same meaning in the two locations. As we learned in Chapter 4, clocks beat more slowly in the lower ship, for two reasons. Einstein concluded from his Principle of Equivalence that *stationary* clocks run more slowly at a place where gravity is stronger, as it is nearer the Earth, and this prediction has been verified experimentally with great precision. Then, as we know from special relativity, *moving* clocks slow down. The lower ship, which is moving more

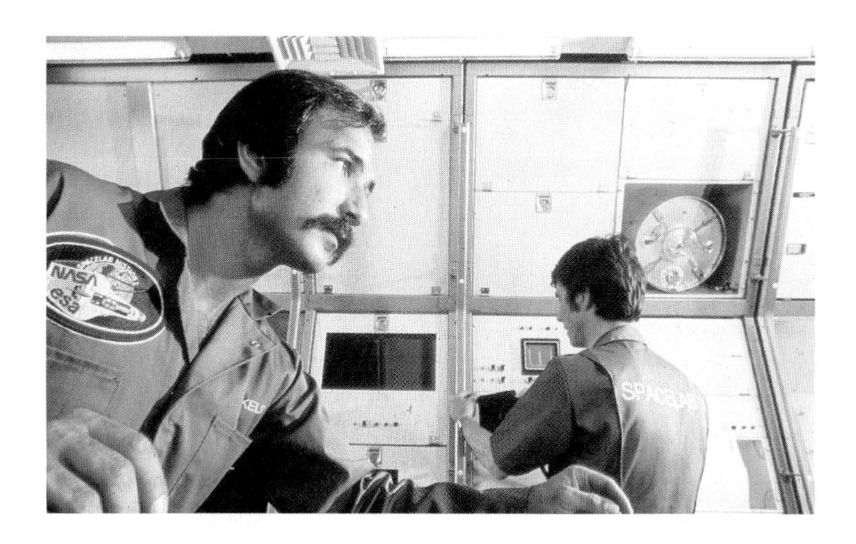

Payload specialists in Spacelab

rapidly in the stronger gravitational field, experiences a greater slowing of its clocks. This additional special-relativistic effect is half of the gravitational one.

To put it more generally: Flat geometry applies within small regions, but quite distinct reference frames must be used in significantly different locations. A change of one unit in a coordinate does not have the same metrical significance at different locations. Does that ring a bell? Yes, the message is clear: the geometry of space-time in the presence of a gravitational field is *curved*.

Einstein

Albert Einstein left Zürich in 1911 to become professor of theoretical physics at the German University of Prague, which was founded in 1348 by the Holy Roman Emperor Charles IV. This was a short-lived move. By the summer of 1912, he was back in Zürich, returning in triumph, as professor of theoretical physics, to his old school, the Polytechnic, the Swiss Federal Institute of Technology. There (happy circumstance!) he again met his old friend Marcel Grossmann, now a professor of mathematics at the same institute. Why was it a happy circumstance? Because by now Einstein had gotten the message:

> It could no longer be required that coordinates should signify direct results of measurement. I was much bothered by this piece of knowledge, for it took me a long time to see what coordinates in general really meant in physics. I did not find the way out of this dilemma till 1912.[5]

He had recognized the necessity for a four-dimensional Riemannian metric. About this, Einstein would later remark:

> I first learned of the work of Riemann at a time when the basic principles of the general theory of relativity had already long been clearly conceived.[6]

Now he needed to translate these basic principles, incorporating the physical concepts of gravitation and the insight of the Principle of Equivalence into an unfamiliar mathematical language. Grossmann was there to guide him through the mathematical literature; in the intervening sixty years, Riemann's ideas had been elaborated by several German and Italian geometers. At the beginning of this activity, Einstein wrote to a colleague:

> I am now exclusively occupied with the problem of gravitation and hope, with the help of a local mathematician friend, to overcome all the difficulties. One thing is certain, however, that never in life have I been quite so tormented. A great respect for mathematics has been instilled within me, the subtler aspects of which, in my stupidity, I regarded until now as pure luxury. Against this problem the original problem of the theory of relativity is child's play.[7]

During the two years of this collaboration, Einstein and Grossmann followed a tortuous and tortured path. Like true love, the course of discovery never did run smooth. Einstein said, at the time, about this cooperative effort:

A colorful mixture of physical and mathematical requirements has been employed as a heuristic device in the investigation.[8]

Later, looking back at it, he remarked,

Already two years before the final publication of the general theory of relativity, we had considered the correct field equations of gravitation, but we failed to recognize that they were physically applicable.[9]

It's not implausible to bifurcate the "we" of this statement: Grossmann, the mathematician, suggested the use of Riemann's curvature; Einstein, the physicist, still new at this game, rejected it.

The collaboration ceased in the spring of 1914, when Einstein accepted an appointment in Berlin as the director of the Kaiser Wilhelm Institute, the research institute of the Prussian Academy. About this, Einstein, a determined Swiss citizen, said,

The Germans are betting on me as a prize hen; I am myself not sure whether I am going to lay another egg.

He did: a *golden* egg.

Matter and Metric

In the presence of a gravitational field, the geometry of space-time, as described by its metric, is curved. This raises two basic questions. Given the metric, how does matter move in that geometry? And, given the distribution of matter, which produces the gravitational field, what is the space-time metric? These are the two sides of the challenging palindrome: *matter* in *motion* produces a gravitational field, which determines the space-time metric that controls the *motion* of *matter*.

In the small region within a freely falling spaceship, matter, which now includes light, moves in straight lines and at constant speeds. These are the straight lines of a *four*-dimensional space; not only is a change in one spatial coordinate proportional to a change in any other, but it is also proportional to a change in the time.

And, now, by connecting together the straight lines within such small regions, we produce a straight line in the larger sense, a *geodesic* of the given geometry. That geodesic describes the *motion* of *matter.*

In the next step, it helps to use units of space and time such that c, the speed of light, is equal to one. Astronomers do it; in *one* year, light goes a distance of *one* light-year. Here's another example: the unit of time is one microsecond, the unit of distance is 299.79 meters. With such units, energy and mass are one and the same: $E = m$.

Warning: The Surgeon General Has Determined That the Next Seven Paragraphs Are Dangerous

Gravitational Field Equations

In the Newtonian description of gravitation as a property of matter, all that counts is the distribution of mass in space, the *density of matter,* which at a given point is specified by one number. But matter, acted on by forces, is in motion. Its flow, or flux, is measured by the product of density and velocity at each point. Because there are three independent directions for that flow, three additional numbers must be specified at each point, making four in all, to describe the density and flow of matter. Is that the whole story? Not according to special relativity. Mass is energy, and energy and momentum are partners. As described in Chapter 3, the shift to a relatively moving reference frame mixes momentum with energy. We must include the *density of momentum,* which is specified by three quantities, and the *flux of momentum,* which is specified by nine quantities because there are three possible directions of flow for each of the three directions of momentum. All this would add up to $1 + 3 + 3 + 9 = 16$ quantities, if they were all independent.

A number of these quantities are identical, however. The flux of energy, *energy* density multiplied by velocity, is the same as the density of momentum, *mass* density multiplied by velocity; this gives three equalities. Furthermore, the flux of x-momentum in the y-direction, for example, which is the product of mass density, x-velocity, and y-velocity, is the same as the flux of y-momentum in the x-direction. This, together with the similar xz and yz identities, gives three more equalities—six in all. And so the number of quantities at each point that are needed to specify the density and the flux of energy and momentum is reduced from sixteen to ten—the mystical tetractys of the Pythagoreans but, more important, just the number of quantities that specify the metric of four-dimensional space-time.

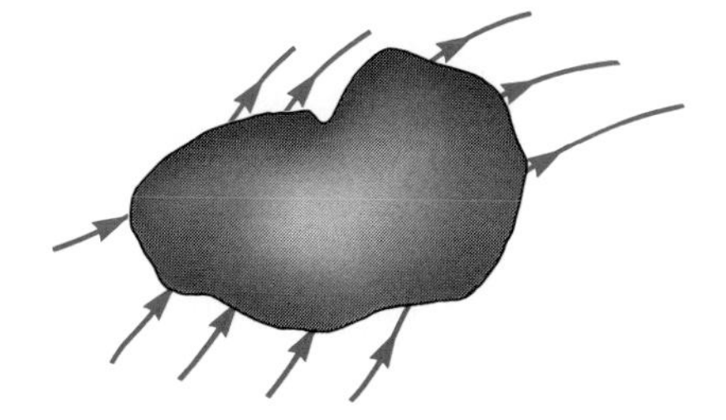

Energy and momentum are conserved properties. Does that not place some restrictions on the ten quantities that describe their spatial distribution and flux? Indeed. Let us begin in the flat space-time of special relativity. The material *energy* in a certain three-dimensional volume changes only because there is an energy *flux* across the two-dimensional boundary surface of that volume; energy is neither created nor destroyed within the volume. For a very small volume, this becomes a local (referring to the immediate neighborhood of a point) relation between the rate of change of energy *density* with time and the rate of change of energy *flux* with position in space. Adding the three analogous relations for momentum gives *four* restricting relations in all. There are four similar but somewhat more complicated relations in a curved space-time—what is different there is that the energy and momentum of matter are no longer conserved; these material properties are exchanged with those of the gravitational field.

Consider an apparently similar circumstance—that of electrically charged particles in the presence of an electromagnetic field, where the mechanical properties of the particles are also not conserved, being exchanged with those of the electromagnetic field. Here one can supplement the various mechanical densities and fluxes with those of the electromagnetic field and recover the four simple, local conservation statements about the whole system. Ah, but there is something different about a gravitational field. *You* tell me that the density of gravitational energy at a certain point is so much. *I* tell you that it is *zero*—because I am in a frame of reference that is falling freely at that point, and the Principle of Equivalence insists that all sign of gravitation is thereby removed. Of course, I can do that only in a small region. The reality of gravitational energy in the large is not questioned; it is just that it is impossible to describe it uniquely at individual points. This is the crux of the torment that Einstein experienced in trying to reconcile his preconceived physical principles with the mathematics of curved space-time.

Matter in motion generates a gravitational field, which produces the curvature of the space-time geometry. To give a precise statement of this connection between geometry and physics, we look for a relation between the measures of curvature and the measures of matter. As we know, the description of moving matter requires that ten quantities be specified at each point; we call them, collectively, T. How do we find ten analogous quantities that specify curvature? Here's how to do it.

At each point there is a mean curvature that depends on the metric but is independent of the particular choice of coordinates. Multiply this number by the volume of a small four-dimensional region, which volume is also independent of the choice of coordinates. The sum of all such products, covering the entire space, is a kind of total curvature; it depends on the overall metric but not on the particular coordinate system used to convey the metric. We then ask about the *change* in the total curvature when the metric is altered by small, arbitrarily chosen amounts at each point. That change is also a sum of products. The factors of an individual product are: the volume of a small region enclosing a point; the change in one of the ten components of the metric at that point; a multiplicative coefficient corresponding to this component. After adding together the ten possible changes at a point, we proceed to cover the entire space by adding the changes within all such small volumes; the result is the change in the total curvature. The ten *coefficients* at a point, which depend on the metric, are the sought-for curvatures. They constitute the *Einstein curvature,* which we designate by the symbol G.

Now comes the decisive step. Suppose that we change the coordinate system a little bit, as expressed by small, arbitrarily chosen modifications in each of the four coordinates. Such a coordinate change produces small changes in the metric. But the coordinate-independent total curvature must *not* be altered. Therefore, at each point, with its arbitrary changes in the four coordinates, there are four restrictive conditions on G. They turn out to be *identical* in form with the four restrictive conditions on T! Immediate conclusion: G is proportional to T; that is, each of the ten measures of curvature associated with the gravitational field is proportional (with a universal constant) to the corresponding measure of the matter that produces the gravitational field. These are Einstein's equations for the gravitational field.

Triumph

Early in 1915, Einstein began to act as a physicist again; he thought about the possible *experimental* consequences of his work. This seems to have set him on a new track, for, in three consecutive weekly meetings of the Prussian Academy, in November of that year, he reported:

During the past years, I have tried hard to construct a general relativity theory. . . . I actually thought that I had discovered the only law of gravitation. . . . I lost all faith in the field equations. . . . I reverted to the requirement of the [unrestricted coordinate choice] of the field equations, from which I had departed with a heavy heart . . . three years ago when I worked together with my friend Grossmann. . . At that time we had already come quite close to the solution. . . The general relativity theory is finally completed as a logical structure.

In its most general formulation, which turns the space-time coordinates into parameters devoid of physical significance, the principle of relativity leads, with conclusive necessity, to quite a definite theory of gravitation which can explain the perihelion motion of Mercury."[10]*

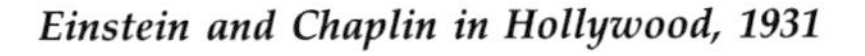

Einstein and Chaplin in Hollywood, 1931

*How different is the real story of these gropings, false trails, and torments from the Hollywood-type versions. Charles Chaplin, in his autobiography, writes of a dinner in 1926 at which Mrs. Einstein told of the morning Einstein conceived the theory of relativity. After hardly touching his breakfast, he said, "I have a wonderful idea." He went upstairs to his study, stayed there for two weeks, and then came downstairs, put two sheets of paper on the table, and said, "That's it."

The great Göttingen mathematician David Hilbert, who had heard Einstein speak about his earlier attempts, published the general equations of gravitation almost simultaneously with Einstein. Although this work was quite independent, Hilbert never denied that the inspiration was Einstein's. He once said, "Every boy in the streets of Göttingen understands more about four-dimensional geometry than Einstein. Yet, in spite of that, Einstein did the work and not the mathematicians." Hilbert even recommended that Einstein be awarded the mathematics prize named for Bolyai. (Einstein did not receive the prize.)

Newton and Einstein

What was that about the perihelion motion of Mercury? And what is the story of the gravitational deflection of light by the Sun, in limbo since Chapter 4? Both questions set the stage for a confrontation between the spirit of Newton, wielding forces and accelerations, and that of Einstein, with his curvatures and geodesics. How different these languages are, and yet how similar their physics must be, at least within the solar system, where Newton's laws have been the infallible rule book of space navigation.

To appreciate the way in which Einstein's physics includes and extends Newton's physics, think of T, the ten measures of matter in motion, as contrasted with the single measure of Newton, the density of matter. The other nine components of T refer explicitly to motion in terms of speed relative to the speed of light (recall that we are using units such that $c = 1$). On this scale, solar-system speeds are very small; the Earth, for example, moves in its orbit at one-ten thousandth of the speed of light. If we simply ignore that motion, the Einsteinian description should reduce to Newton's, and so it does.

THE DEFLECTION OF LIGHT

It is the dependence of motion, then, that distinguishes Einstein from Newton. To test this conception, we look for such effects in the properties of a body with the highest possible speed. That body is light itself. Einstein's use of the Principle of Equivalence, in 1911, to predict the deflection of light in a grazing passage by the Sun was discussed in Chapter 4. The predicted deflection angle was the same as the Newtonian prediction for a body that happens to have the speed c. We can now appreciate that there should be an additional motional effect, which, for motion at the speed of light, might be comparable in size to the Newtonian effect.

To be more precise, let us recall that the ten quantities composing T are given by the density and flux of energy and of momentum. It is not only energy (mass) that gravitates, but also momentum. Energy and momentum are equal ($c = 1$) for a beam of light, which (correctly) suggests that the prediction of general relativity is *twice** the Newtonian value. That prediction, doubling the 0.875 arc sec of relation 4.2, is 1.75 arc sec, a shift in direction that, at the distance of the Moon, is a displacement of a few kilometers, and at the distance of the Sun is somewhat more than a thousand kilometers (the radius of the Sun is 700,000 km).

Eddington

Einstein published this prediction in 1915, at a time when Europe was convulsed in the suicidal mania of World War I. A single copy of his paper, passing through neutral Holland, came into the hands of the Plumian Professor of Astronomy at Cambridge, Sir Arthur Stanley Eddington. As a conscientious objector to the war, Eddington had the time to appreciate this bit of German physics. Einstein had already pointed out in 1911 that a total eclipse of the Sun was needed to observe the bending of light. But not any eclipse would do. As Eddington explained:

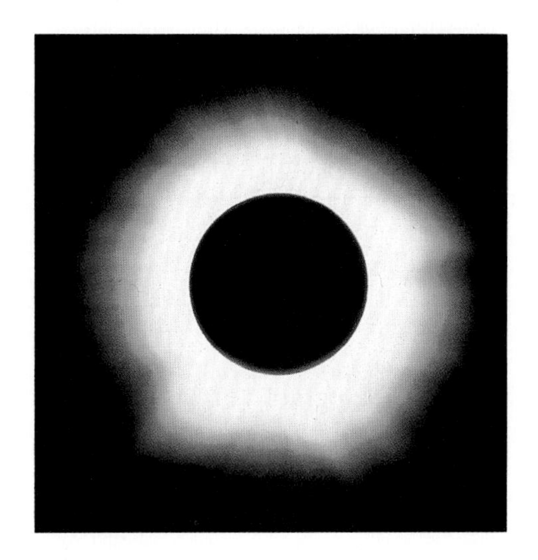

Solar corona

> The bending affects stars seen near the Sun, and accordingly the only chance of making the observation is during a total eclipse when the Moon cuts off the dazzling light. Even then there is a great deal of light from the Sun's corona which stretches far above the disc. It is thus necessary to have rather bright stars near the Sun, which will not be lost in the glare of the corona. Further the displacements of these stars can only be measured relative to other stars, preferably more distant from the Sun and less displaced; we therefore need a reasonable number of outer bright stars to serve as reference points.

In a superstitious age a natural philosopher wishing to perform an important experiment would consult an astrologer to ascertain an auspicious

*This physical approach to light deflection will be reinforced in Chapter 6 by a geometrical discussion.

Sir Arthur Stanley Eddington (1882–1944)

moment for the trial. With better reason, an astronomer today consulting the stars would announce that the most favourable day of the year for weighing light is May 29. The reason is that the Sun in its annual journey around the ecliptic goes through fields of stars of varying richness, but on May 29 it is in the midst of a quite exceptional patch of bright stars—part of the Hyades—by far the best star-field encountered. . . . by strange good fortune an eclipse did happen on May 29, 1919.[11]

The Hyades, or "rain makers" (to the Greeks, the rising of the Sun in this star group signalled the beginning of the rainy season), is a V-shaped cluster of stars in the constellation Taurus. Nearby in the sky, although not part of the cluster, is the bright star Aldebaran. In 1917, the Astronomer Royal, Sir Frank Dyson, called attention to this exceptional opportunity, and two British expeditions were mounted to observe the eclipse at different locations along the line of totality of the eclipse path. One went to Sobral in North Brazil; the other, which included Eddington, set up camp on a race course in the island of Principe, off the coast of West Africa.

There was a fierce thunderstorm on Principe the morning of the 29th. Yet, by the time the five minutes of totality arrived, the sky was clear enough to expose a number of photographic plates. Alas, only two had measurable star images. At first, things seemed much better at Sobral. Although the sky began overcast, it cleared for five of the six minutes of totality there, and twenty-six plates were exposed; twenty-two of them had star images. Unhappily, it turned out that only seven were usable; the Sun's rays had heated and distorted the mirror used to hold the Sun's image steady in one of the telescopes.

To discover a shift in position of stars produced by the proximity of the Sun, it is necessary to have photographs of the same area of the sky at another time of the year, when the Sun is elsewhere. The Sobral team returned in July for that purpose. But Eddington had with him the results of measurements that had been made earlier. Accordingly, he set to work with one of his good plates to measure the positions of the five stars that were visible in proximity to the Sun. Thus far, we have considered only the deflection of light that grazes the Sun, but no star image was actually observed at grazing incidence. The light paths had various distances of closest approach to the Sun, but none was less than twice the Sun's radius. Now, the deflection of light should vary inversely with the

BOX 5·3 ***The Eclipse of 1780***

In planning an eclipse expedition to test Einstein's prediction, while war with Germany still raged, the British were demonstrating again that science takes precedence over nationalistic attitudes. There had been an earlier eclipse; then, the eclipse expedition was mounted by the rebellious colonists of North America. The first total eclipse that they would have had the chance to observe was predicted for 27 October 1780. The only sites on the calculated path of totality that were accessible by water (land transport of the instruments was unfeasible) happened to be at the western end of Penobscot Bay, on the shores of what is now the state of Maine. Unfortunately, the British had just occupied that region.

John Hancock, the first to sign the Declaration of Independence, and, at the time, speaker of the Massachusetts legislature, wrote to the British commander at Penobscot Bay, requesting safe passage:

> Sr. It is expected that there will be a very remarkable Eclipse of the Sun on ye 27th of Octb. next; . . . the Centre of ye Moon's Shadow if the longitude & latitude of that place thru ye Maps can be depended on, being by culmination to pass over Penobscot Bay. . . . The Genl. Assembly of this State have made provision for suitable persons to observe it . . . it is not doubted that as a Friend of Science, you will not only give him yr. permission for that purpose, but every assistance in your power to render the observations as perfect as possible. Though we are politically enemies, yet with regard to Science it is presumable we shall not dissent from the practice of all civilized people in promoting it either in conjunction or separately as occasions for it shall happen to offer.

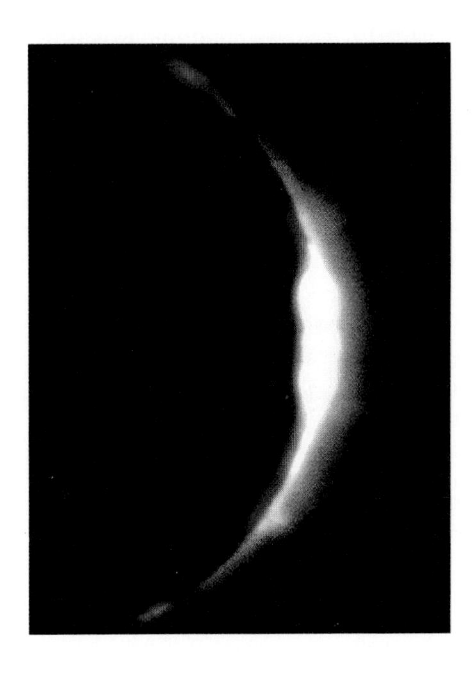

Permission was granted. The person referred to in the letter, the leader of the expedition, was Reverend Samuel Williams, Hollis Professor of Mathematics and Natural Philosophy at Harvard University, then in its 144th year. On the appointed day, viewing conditions were excellent, but it became evident that, perhaps through map errors or miscalculation, they had just missed totality. Yet it was not all wasted. The account of the expedition gave the first description of a phenomenon that would be rediscovered about sixty years later: Baily's beads, the string of small spots of light seen at the Moon's edge just before and just after the period of totality—the light of the Sun penetrating through the valleys of the Moon.

distance from the center of the Sun, at closest approach.* So a measurement for a greater distance can be converted into one of grazing incidence by multiplying the measured deflection by the ratio of the actual distance to the Sun's radius. (If the deflection has a certain value at twice the Sun's radius, it is double that at grazing incidence.) In this way, the five measurements could be compared directly. When Eddington did all this, he found an average shift that agreed with the prediction of general relativity. He later said, "That was the greatest moment of my life!"

Concerning the final results, and the uncertainties of the various measurements, Eddington wrote:

> The evidence of the Principe plates is just about sufficient to rule out the possibility of the "half-deflection" [the 1911 prediction of 0.875 arc sec], and the Sobral plates exclude it with practical certainty.[12]

At the presentation of the expedition's findings to the Royal Astronomical Society on November 6, 1919, the Astronomer Royal remarked, about the images on the plates:

> I am prepared to say that there can be no doubt that they confirm Einstein's prediction. A very definite result has been obtained that light is deflected in accordance with Einstein's law of gravitation.

Eddington informed Einstein, by cable, of this experimental confirmation. According to an eyewitness account, Einstein unemotionally remarked, "I knew the theory was correct." When pressed about his reaction had it gone otherwise, he replied, "I would have felt sorry for the good Lord! The theory is, of course, all right."[13] If you consider his behavior to be arrogant, remember that Einstein already knew of a more fundamental confirmation of his theory in the perihelion motion of Mercury (soon to be discussed).

---

*We know that the general relativistic deflection angle is $4g'R/c^2$ (compare relation 4.1) for grazing incidence. That is, R is assumed to be the radius of the Sun and g' the acceleration of gravity at the surface of the Sun. The same result applies, however, for any larger value of R, with g' correspondingly reduced in accordance with Newton's inverse square law, proportional to $1/R^2$. The net dependence on R is as $1/R$. (Similar statements were made about the gravitational red shift in Chapter 4.)

Radio Telescopes

Such eclipse measurements have continued through the years, but Einstein's prediction was not verified with any great accuracy until the advent of radio astronomy. Radio telescopes used in pairs can measure small angle changes with great precision. The technique uses the same property of waves—interference—on which the Michelson-Morley experiment depended. Here is the basic idea.

Suppose, first, that waves of a definite wavelength, from a radio source in the sky, are received by two radio telescopes in a direction perpendicular to the base line that connects the telescopes. The received signals are relayed to the middle point of the base line, where they are combined. Everything about these two signals being identical, they reinforce each other to produce a stronger signal. Next, suppose that the waves arrive at a slight angle to the perpendicular direction. Then the distance to one telescope is greater than the distance to the other, by an amount that is equal to the length of the base line multiplied by the radian measure of that angle. If this excess distance is half a wavelength—the distance from crest to trough—the two signals, arriving at the midway point, cancel each other.

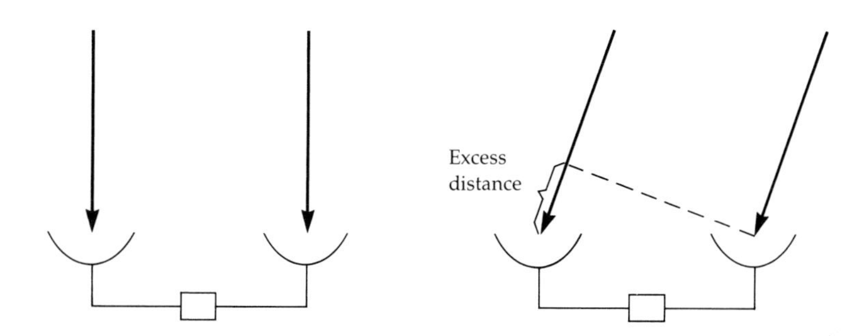

To see how sensitive this technique is to small angle changes, let the radio frequency be 8,085 MHz, corresponding to a wavelength of 3.708 cm, and choose the length of the base line to be 35 km. Then the change of direction that turns reinforcement of the waves into cancellation—the angle in radians equal to the ratio of half a wavelength to the length of the base line—is about a tenth of an arc sec.* Measurements to within a small fraction of this angle are possible.

At the National Radio Astronomical Observatory in Green Bank, West Virginia, just such a base line exists between three large antennas, separated by a few kilometers, and a somewhat smaller one high in the Alleghany Mountains. The remote telescope can be paired with each of the other three. The antennas receive at the 8,085-MHz frequency, and at 2,695 MHz. With that information, the frequency-dependent effect on the propagation of radio waves that is produced by the electrically charged particles in the hot solar corona (more significant for radio waves than for light) can be removed.

Nature has been very cooperative in providing strong radio sources in the sky, so positioned that the Sun passes near them, as seen from the Earth. In April, the Sun moves through such a group

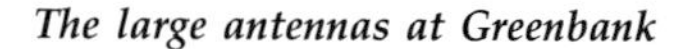

The large antennas at Greenbank

*This angle is

$$\frac{1}{2}\left(\frac{3.708 \times 10^{-5}\ \text{km}}{35\ \text{km}}\right) = 0.530 \times 10^{-6}\ \text{rad},$$

or, because 1 rad = 2.06×10^{5} arc sec, it is

$$0.530 \times 0.206 = 0.109\ \text{arc sec}.$$

of sources: three quasars (quasi-stellar objects) that are located nearly in a straight line in the sky. Deflection measurements made in Greenbank in 1975 have confirmed Einstein's prediction to within one part in a hundred.

THE PERIHELION MOTION OF MERCURY

Ceres

The five planets of antiquity (to which Copernicus added the Earth in the sixteenth century) were joined by Uranus in 1781 (see Box 5.4). Twenty years later, the Swabian German philosopher Georg Wilhelm Friedrich Hegel (1770–1831) declared that *seven* was just the right value for the total number of planets. Even before that statement was published, reality had proved otherwise.

It had been known since 1766 that the radii of the planetary orbits have a striking regularity, which is called Bode's law (it was

BOX 5·4 ***Uranus and Neptune***

The planet Uranus had been seen on a number of occasions, beginning in 1690, but not until 1781 was that celestial object recognized as a planet (by Sir William Hershel). In 1820, the suspicion arose that the motion of Uranus did not conform to Newton's grand scheme, even when the gravitational disturbances produced by the massive inner planets, Jupiter and Saturn, were included. That the solution could be found in the influence of an unknown outer planet was suspected, but where should one look for that faint, moving intruder? In 1845, the answer was given by John Couch Adams (1819–1892), a fellow of St. John's College, Cambridge University, and somewhat later but independently by Urbain Jean Joseph Leverrier (1811–1877). It took the agreement of the two predictions to get the attention of the observational astronomers. On the nights of September 23 and 24, 1846, at the Berlin Observatory, the predicted planet was discovered. The Copley medal of the Royal Society was awarded to Adams in 1848, which was also the year that the Adams Prize was founded. As mentioned in Chapter 1, Maxwell won it eleven years later on the subject of Saturn's rings.

discovered by Johann Daniel Titius). Consider this sequence of numbers:

0, 3, 6, 12, 24, 48, 96, 192, ... ,

in which each successive number, after the first, is double the preceding one. Add four to every number to produce the following sequence:

4, 7, 10, 16, 28, 52, 100, 196, ...

Now compare this sequence with the succession of radii of the planetary orbits, each taken to be half the longest dimension of the elliptic path. (For convenience, divide each set of numbers by its third member, thereby reducing that entry to one.) The comparison appears in Table 5.1, with the outer planets displaced one step.

TABLE 5·1 ***Comparison of the Titius-Bode Number with Relative Planetary Radii***

Titius number	0.4	0.7	1.0	1.6	2.8	5.2	10	19.6
Relative radius	0.3871	0.7233	1.0	1.524		5.203	9.539	19.18
Planet	Mercury	Venus	Earth	Mars		Jupiter	Saturn	Uranus

The gap thus disclosed between Mars and Jupiter inspired a search for the "missing" planet. Meanwhile, normal astronomical work continued. In Palermo, Giuseppi Piazzi (1746–1826) was preparing a star catalog. On the first day of the nineteenth century (1 January 1801), he found something new. By the third night, this object had *moved;* it was a comet or a *planet.* Piazzi made observations for forty-two nights, after which illness forced him to stop. He had, however, written a letter to alert Bode [Johann Elert Bode (1747–1826)] in Berlin. Unfortunately, by the time the letter reached Bode, in late March, the moving object had been lost in the glare of the Sun.

There was the problem: given Piazzi's observations during a time interval that was far too short for the then-current computational methods, was it possible, nevertheless, to predict the path of the object so that it could be rediscovered once it had cleared the neighborhood of the Sun? Gauss accepted that formidable challenge.

How much information should be sufficient to predict the orbit of a body, according to Newton's laws? It is enough to know the position and the velocity of the body at the same time. Then Newton's laws supply the acceleration at that time, from which the rate of change of acceleration can be inferred, and so on. Step by step, the orbit evolves. Position is a directed quantity in three-dimensional space. Three numbers, its coordinates, are needed to specify a position. Similarly, three numbers are needed to specify a velocity. Thus, six numbers in all are required to predict an orbit. Now, an astronomical observation at a certain time is the measurement of the direction of a distant body, as specified by two angles analogous to latitude and longitude. Observations at *three* different times therefore supply *six* pieces of information, which, indeed, are enough to predict the whole orbit.

By November 1801, Gauss had completed his calculations (no small feat, a century and a half before the computer age); the new object was rediscovered (reacquisitioned, in modern parlance) on the night of December 31, 1801. It was the first of the minor planets, or asteroids: Ceres, so named for the patron deity of Sicily. The radius of Ceres's orbit is remarkably close to that expected according to Bode's law. But tiny Ceres was *not* the sought-for major planet. By now, thousands of such minor planets are known.

Mercury

Newton's laws were taken for granted in Gauss's work. Now, more than a century later, the focus is on deviations from the Newtonian laws. Light-deflection experiments use the ultrahigh speed of light to make the motional effects of general relativity as large as possible. But there is another way. Much smaller effects can also be detected if their consequences accumulate in time. This brings to mind the periodic orbits of planets, repeatedly swinging around the Sun. What can happen there?

It helps to think of an extreme example: a highly eccentric elliptical orbit in which the planet plunges close to the Sun and then

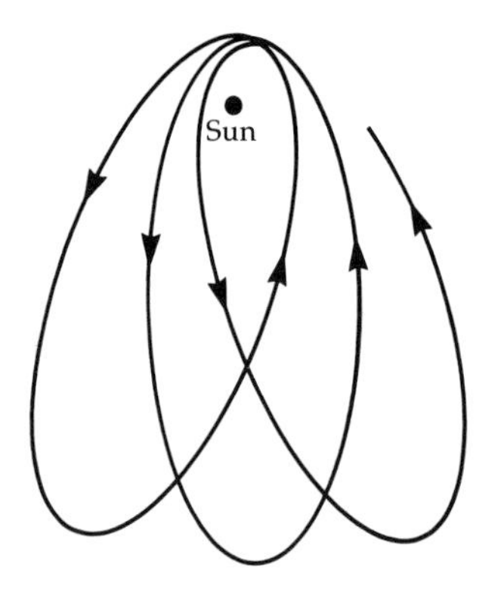

moves far out in space again before returning on the identical path—at least a path that, according to Newton, would be identical. In fact, as the planet falls in toward the Sun it speeds up, and the motional effects of general relativity come more into play, *increasing* the force between the planet and the Sun. As a result, the planet races a bit farther around the Sun than a Newtonian one would before embarking on its more leisurely trip out to the point of return. Consequently, the axis (the long dimension) of the planetary orbit slowly turns in the same sense as the orbital motion. That turning, or *precession,* is most noticeable at perihelion, the orbital point nearest the Sun, and it will be the larger the more one realizes this extreme example.

This picks out the planet Mercury, which is closest to the Sun and has the most eccentric of the planetary orbits, apart from that of remote, frigid Pluto. Since the time of Leverrier, who, like Adams, had predicted the existence of the planet Neptune from the orbital irregularities of Uranus (see Box 5.4), it had been known that the axis of Mercury's orbit rotates 574 seconds of arc per century. Of this amount, 531 seconds could be accounted for by the gravitational disturbances produced by the known planets. In anticipation of a triumph similar to the discovery of Neptune, the name Vulcan was already chosen for a planet, interior to Mercury, that would account for the remaining 43 seconds per century. (By itself, this effect would need 30,000 centuries to turn the axis full circle.) Fiery Vulcan was searched for, but it never materialized. So it was a stunning victory for general relativity that the additional precession it predicted was precisely that 43 seconds of arc per century.

In a letter written toward the end of 1915, Einstein, looking back at the tortured path he had followed, exulted, "The final release from misery has been obtained. What pleased me most is the agreement with the perihelion motion of Mercury."[14]

The perihelion effect involves more of general relativity than does the deflection of light. As discussed in Chapter 4 in connection with the Principle of Equivalence, it is the essence of general relativity that all forms of energy including gravitational energy, gravitate. To a certain accuracy (see Box 5.5), the good agreement of the perihelion effect shows that the gravitational energy between the planet and the Sun is also acted on gravitationally by the Sun.

Mariner 10 photograph of Mercury at a distance of 200,000 km

BOX 5·5 *Perihelion Precession*

The motional effects of general relativity produce deviations from Newton's laws—in particular, the slow turning of a planetary elliptic orbit known as the precession of the perihelion. In the solar system, speeds are small on the scale set by c. A suggestion of what is significant comes from the relativistic factor gamma, which, for small values of the speed ratio v/c, differs from unity by $\frac{1}{2}(v/c)^2$. In fact, multiplying this difference by 6, to get $3(v/c)^2$, gives the fraction of 360 degrees through which the orbit turns in one revolution, for nearly circular orbits. More generally, one must use the average value of $3(v/c)^2$ and multiply it by the square of the ratio between a, half the longest dimension of the ellipse, and b, half the shortest dimension. For Earth, with a very nearly circular orbit, the value of v/c is 0.994×10^{-4}. The angle of precession in 100 Earth revolutions, one century, expressed in seconds of arc (note that $360° = 360 \times 60 \times 60 = 1.296 \times 10^6$ arc sec) is

$$10^2 \times 3\,(0.994 \times 10^{-4})^2 \times (1.296 \times 10^6) = 3.84 \text{ arc sec/century.}$$

That is hard to measure, not only because it is so small, but also because Earth's orbit is so nearly circular that the point of perihelion is difficult to locate. (If the orbit were exactly circular, how would you know it was precessing at all?)

The innermost planet Mercury is more favorable. We can work out its precession from that of Earth, basically by knowing the ratio of the value of a for Mercury to that of Earth, 0.3871 (see Table 5.1). According to Kepler's third law, the cube of a is proportional to the square of the orbital period T. First, this says that the average value of $(v/c)^2$, which is proportional to a^2/T^2, varies as $1/a$. Then, in computing the rate of precession per century, we need the number of revolutions in that time interval, which is found by dividing 100 years by the period T. Now, T is proportional to $a^{3/2} = a \times \sqrt{a}$. Therefore the rate of precession per century varies as the inverse of $a^2 \times \sqrt{a}$. This alone says that Mercury's precession rate exceeds that of Earth by the factor $(1/0.3871)^2 \times 1/\sqrt{0.3871} = 10.73$. Then there is the small effect produced by Mercury's value of $(a/b)^2 = 1.044$. The product of the two numbers equals 11.2, and the precession of Mercury is predicted as

$$11.2 \times 3.84 = 43.0 \text{ arc sec/century.}$$

The combination of optical and radio measurements of the planets has produced an experimental value that agrees with this to better than 1 percent. One-sixth of the theoretical value is produced by the gravitational action of the Sun on the gravitational energy between the planet and the Sun. The agreement between theory and experiment shows that gravitational energy gravitates, to a precision about that of the Principle of Equivalence test discussed in Chapter 4.

Nevertheless, it was the light-deflection findings of the British eclipse expedition of 1919 that caught the attention of the world. The immediate and continuing public adulation of Einstein is familiar to all. But from the very beginning, Einstein retained his sense of proportion and sense of humor. A letter to the *London Times* contained, as he put it, this additional application of the principle of relativity: "Today I am described in Germany as a 'German savant,' and in England as a 'Swiss Jew.' Should it ever be my fate to be represented as a bête noire, I should, on the contrary, become a 'Swiss Jew' for the Germans and a 'German savant' for the English."

NOTES

1. **George Sarton:** *A History of Science* (Harvard University Press, 1952), p. 115.
2. **J. Mehra, ed.:** *Physicist's Conception of Nature* (D. Reidel, 1973), p. 111.
3. **Sarton:** *History of Science*, p. 504.
4. **H. Weyl:** *Space-Time-Matter* (Dover, 1950), p. 97.
5. **A. Einstein:** *Essays in Science* (Philosophical Library), p. 81.
6. **H. Dukas and B. Hoffman, eds.:** *Albert Einstein: The Human Side* (Princeton University Press, 1979), p. 19.
7. **Mehra:** *Physicist's Conception of Nature*, p. 93.
8. Ibid., p. 155.
9. Ibid., p. 154.
10. Ibid., pp. 106 and 108.

11. Quoted in **J. Newman, ed.:** *World of Mathematics* (Simon and Schuster, 1956), p. 1100.
12. Ibid., p. 1101.
13. **Mehra:** *Physicist's Conception of Nature,* p. 131.
14. Ibid., p. 157.

6

AT THE FRONTIER

The findings of the British eclipse expedition of 1919 were a smashing triumph for general relativity. Three years later, Einstein wrote,

> The scientific theorist is not to be envied. For Nature, or more precisely experiment, is an inexorable and not very friendly judge of his work. It never says "Yes" to a theory. In the most favorable cases it says "Maybe," and in the great majority of cases simply "No." If an experiment agrees with a theory it means for the latter "Maybe," and if it does not agree it means "No." Probably every theory will someday experience its "No"—most theories, soon after conception.[1]

What sort of responses has "Nature, or more precisely experiment," been giving Einstein's general theory of relativity lately? And what is in the offing?

The data available to Einstein in the first decades of this century came from optical astronomy. Now we are in the space age and the time of radio astronomy. The tools of these advanced technologies have sharpened the results of earlier tests and produced new ones. Armed with devices that exploit the special properties of very low temperature matter, other experimenters seek evidence that, through the medium of gravitation, the stars guide us and send us signals of their more violent activities.

Sixteen days after the targeted 200th anniversary date of July 4, 1976, the lander of Viking 1 settled on the surface of Mars, in the region named Chryse. Viking 2, which followed later, had another locale, called Utopia, selected for its lander. The emotional experience of receiving pictures from the surface of another planet is unforgettable. Those television signals were carried from Mars to

Earth by radio waves. And those radio waves were to bring another message, another "Maybe," for Einstein's general theory of relativity.

The Earth–Mars connecting line relative to the Sun

THE SPEED OF LIGHT

At Goldstone Deep Space Station in the Mohave Desert of California, radio waves are used to track spacecraft. A Viking lander sitting on the Martian surface receives the radio signals from Earth and relays them back at the same frequency. The Viking orbiter also does this, and at a different frequency, as well. In empty space, radio waves move at the invariable speed of light, 300,000 km/s. The measured time for a round trip can therefore be converted into the distance between Mars and Earth, at successive points in the orbits of these planets about the Sun. This accurately measured distance changes smoothly in time from a minimum value of about 80 million km to a maximum of 380 million km—except when the path of the radio waves connecting Earth and

Simulation at Jet Propulsion Laboratory of Viking 2 on Mars (facing page, left); a self-portrait of Viking 1 on Mars (facing page, right); Mars seen by the approaching Viking 1 (above, left); the Martian surface seen by Viking 2 (above, right)

Mars passes close to the edge of the Sun. Then unusual things happen. Unusual, but not unexpected.

Space is not entirely empty near the Sun. There is the solar corona, a region of energetic, electrically charged particles, and they influence the speed of electromagnetic waves. Radio waves of different frequencies are affected differently in regions of electrical

The Mars dish at Goldstone

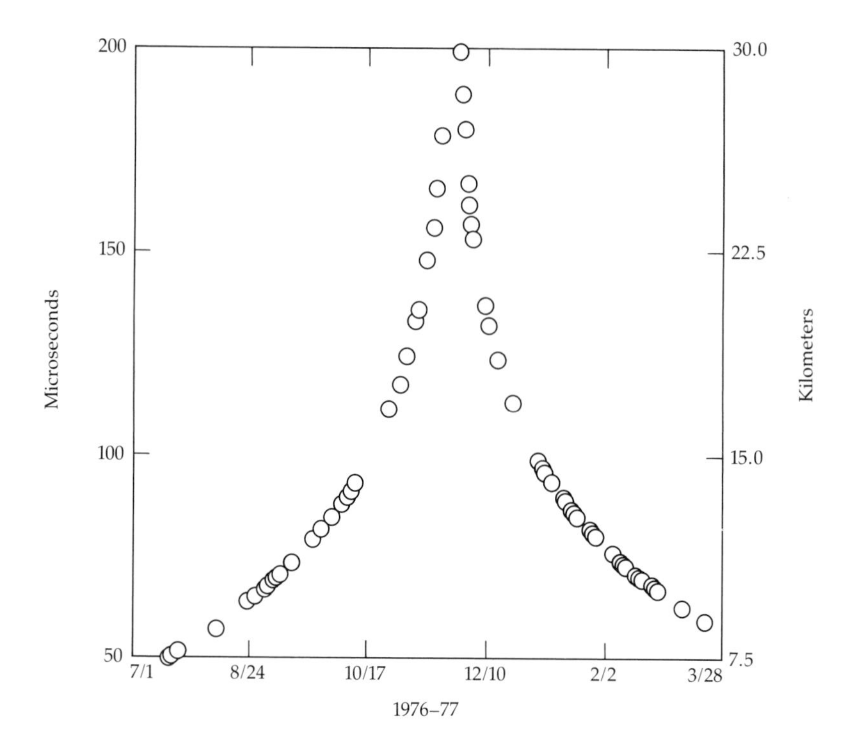

Lengthening of round-trip travel time of radio waves between Earth and Mars, and the equivalent displacement of Mars

activity; the lower the frequency, the stronger the influence. Because Viking speaks to Earth in two voices, the coronal effect can be identified and removed (this was referred to in Chapter 5). And, when that is done, something interesting remains.

As the radio waves approach the edge of the Sun, Mars seems to move out of its orbit, retreating by more than 30 km before returning to its accustomed path. For about two-and-a-half months, Mars appears to be displaced by more than 15 km. Unless you prefer to believe that Mars does swerve out of its orbit, we can conclude from the observed lengthening of the radio travel time that electromagnetic waves passing near the Sun, in space empty of matter, move more *slowly* than usual.

Newton

Light, electromagnetic waves, slowing down in empty space! Is that expected? Well, it depends on to whom you talk. Isaac Newton thought that light was composed of particles. Had anyone asked him about the behavior of light that passes close to the edge of the Sun, he undoubtedly would have replied in proper Newtonian fashion: light should behave like any other particle; it is distinguished only by its very high speed. Being gravitationally attracted by the Sun, it falls toward the Sun and speeds up until it reaches the point of closest approach. Thereafter, receding from the Sun, it slows down and eventually regains its initial speed but is deflected from its original direction. As mentioned in Chapter 4, the Newtonian prediction for light deflection was first worked out early in the nineteenth century (1804).

Well, light *is* deflected by a close encounter with the Sun, but by *twice* the Newtonian prediction. That was the conclusion of the 1919 eclipse expedition, and it has been strengthened by later experiments that use radio waves. But, light speeds up near the Sun? That will *not* do at all.

Einstein (1911)

We do better by talking to the Albert Einstein of 1911. *He* knew that light slows down near the Sun. That is how he worked out his prediction of the deflection of light, which happened to coincide with the Newtonian value. But, as he predicted only half of the observed deflection, he would fail correspondingly to predict the amount of light slowing. Half way measures *will not* do. (It has been said that Einstein did not explicitly remark on this additional test of general relativity. Superficially true—but he thoroughly understood the essential point, the altered speed of light in a gravitational field.)

Einstein (1915)

That brings us to the Albert Einstein of late 1915. He had just crowned his labors of almost a decade with the completed general theory of relativity, which goes beyond the Principle of Equivalence of 1907–1911 in giving a unified treatment of time and space in the presence of gravitating matter. He had worked out the deflection of light by assuming that the light path is a geodesic of the curved metric, one that is of *zero* length, which generalizes to curved space-time the description, in flat space-time, that is given by $L^2 - (cT)^2 = 0$. This Einstein might well respond to our query with the following argument based on an analogy with special relativity.

In a uniform gravitational field that produces the acceleration g', moving down a height h makes clocks run more slowly, by the fraction $g'h/c^2$, as we learned in Chapter 4. Now, we can recognize

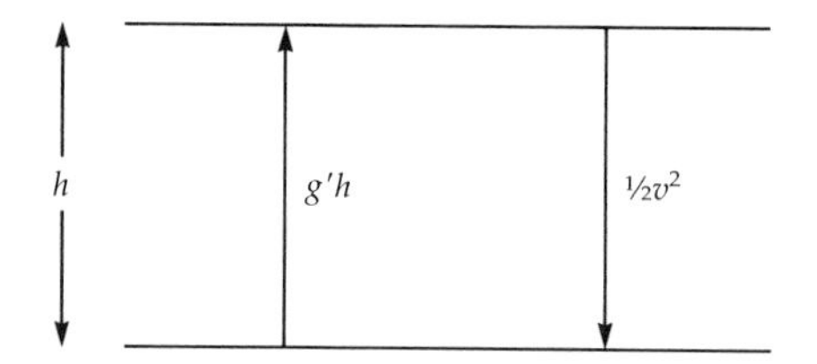

in $g'h$ the potential energy of a unit mass that has been raised a distance h. This quantity is therefore also equal to the kinetic energy, $\frac{1}{2}v^2$, which that unit mass will acquire in falling through height h (assuming that v/c is small compared with unity). So the fractional change in a clock's period is also given by $\frac{1}{2}(v/c)^2$, which is exactly the special relativistic time *dilation* associated with motion at speed v. But, in special-relativity theory, there is another effect associated with motion at speed v: the *contraction* of length in the direction of motion by the same fraction, $\frac{1}{2}(v/c)^2$. This makes it plausible that, on descending a distance h in a uniform gravitational field, length, in the direction of that field, is contracted by the fraction $g'h/c^2$.

The fractional time or space change, $g'h/c^2$, gives the dependence of the measurement standards on position, in a uniform gravitational field of acceleration g'. What we need, however, is the relation between the metric at any point outside the Sun and the metric at very large distances where the observations are performed. That question came up and was answered in Chapter 4 in the discussion about the red shift of the Sun's light. But there the focus was on the surface of the Sun, at radius R, and the fractional red shift was $g'R/c^2$, in which g' is the acceleration of gravity at radius R. There is, however, nothing special about R in this answer; it can be replaced by any larger radius r, provided we remember that the gravitational acceleration varies inversely as the square of r. (This point has been made in Chapters 4 and 5.) For g', then, we substitute $g'(R/r)^2$, which incorporates the inverse square dependence on r of the gravitational acceleration and which reduces to g' at $r = R$, at the solar surface. The result of multiplying this acceleration by r, and dividing by c^2, is $g'(R/r)^2(r/c^2) = g'(R/c)^2/r$. We now see how the fractional dilation in time and the equal fractional contraction of distance in the direction of the gravitational field, the radial direction, approach zero as one recedes from the Sun; they vary inversely with the distance from the Sun.

Space is three-dimensional. What about the scale of distance in the two directions perpendicular to the radial line from the center of the Sun? From what has been said so far, this appears to be unaltered: when the direction of that line is changed by a small angle, the distance covered equals that angle (in radians) multiplied by radius r, the distance to the center of the Sun. Is that the whole story? Not if we remember Einstein's hard-won recognition that the theory "turns the space-time coordinates into parameters devoid of physical significance." The parameter r has no meaning beyond the way in which it enters these calculations of length; we are free to replace it by anything else, although it is convenient to retain its initial interpretation at very large distances. Suppose that we replace r by $r - g'(R/c)^2$, which is a fractional decrease by the amount $g'(R/c)^2/r$. Under the conditions that we are considering, in which the fractional changes are all very much smaller than unity, this shift in r will not alter significantly either the radial contraction factor or the time dilation factor. But now distances in the other two spatial directions are also contracted, by the same factor that applies to the radial direction. And so, with these space-time coordinates, a *universal* fractional change governs time dilation and distance contraction in *any* direction.

A light beam travels at speed c far from gravitating masses. As it nears the Sun, it takes longer (Einstein, 1911) to go a shorter distance (Einstein, 1915). The two fractional changes are equal. Light slows down by *twice* the fractional change that the consideration of time alone predicted. There it is again.

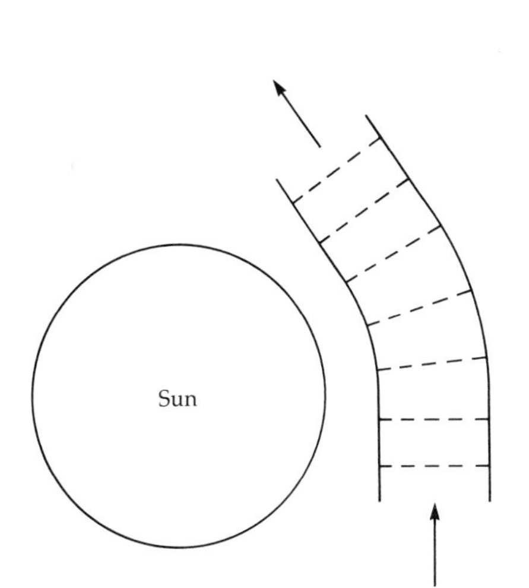

Looking back at Chapter 4, and supplying the additional factor of two, we now find that the angular deflection of light in a grazing passage of the Sun is $4gR/c^2$; the numerical value, an angle measured in radians, is 8.48×10^{-6}. A grazing passage is one for which the distance of closest approach is R. Should the closest distance be a larger value, a, the deflection angle will be smaller by the factor R/a; it is then $4g'(R/c)^2/a$. That angle measures how, in a given time, the travel distance past the point of closest approach changes with the alteration of a (recall the analogy with the turning of a parade in Chapter 4).

Now, finally, we come to the measure of the apparent distance that radio waves cover between Earth and Mars, as the radio beam nears the Sun. The rate at which this apparent distance changes with a in consequence of the varying speed of light, as has just been described, is the deflection angle: $4g'(R/c)^2/a$. If we express a in units of the solar radius, 0.696×10^6 km, the rate at which the apparent displacement of Mars out of its orbit changes with a is equal to

$$8.48 \times 10^{-6}\frac{R}{a} = \frac{8.48 \times 0.696 \text{ km}}{a \text{ solar radii}}$$

$$= \frac{5.90}{a} \text{ km per solar radius.}$$

The Viking measurements confirm this with an accuracy approaching one part in a thousand. That *will* do.

Light slows down near the Sun, but by no more than about four parts in a million. The effect would be much larger, however, for more-compact stars, such as white dwarfs and neutron stars. It is even conceivable that, close to a completely collapsed star, light would come to a halt! That dramatic example of extreme general relativistic behavior is called a *black hole*. The attempt to prove that such bizarre objects exist is at the frontier of current astronomical research. But that is a story in itself. Along with the implications of general relativity for the structure of the whole universe, such predominantly astronomical aspects of Einstein's legacy are outside the scope of this book.

MACH AND EINSTEIN

Ernst Mach

The name of the Austrian physicist Ernst Mach is not unknown these days: Mach 1, Mach 2—that is how one gives the ratio of the speed of an object to the speed of sound. A S(uper)S(onic)T(ransport), such as the Concorde, is a commercial aircraft that can fly at a speed exceeding Mach 1 (and, preferably, run at a profit).* But the real basis of his fame lies elsewhere. That comes from his early influence on Einstein, and the general theory of relativity.

In 1916, Einstein, looking back at Mach's long life (1838–1916), wrote:

> I say with certainty that the study of Mach . . . has been directly and indirectly a great help in my work. . . . Mach recognized the weak spots of classical mechanics and was not very far from requiring a general theory of relativity half a century ago . . . . It is not improbable that Mach would have discovered the theory of relativity, if, at the time . . . the problem of the constancy of the speed of light had been discussed among physicists.[2]

That was very generous of Einstein.

Mach insisted that all physical statements should refer to observable properties (by which *he* meant primitive sensations such

*Even the Soviets acknowledged this requirement when high operating costs were cited in the official announcement (August 1984) that the pride of Aeroflot, the TU-144 SST, had been permanently removed from service.

as touch, color, and warmth). It was in this spirit that Einstein, in 1905, created the special theory of relativity, by analyzing what it means to synchronize clocks through the use of light signals. Yet, in another paper of that famous year, on Brownian motion, which pointed to observable consequences of the unobserved molecular motion, Einstein was firing a broadside at the physicists who still rejected the atomic idea. Foremost among them was Ernst Mach, who, in criticizing atomic theory, wrote:

> Thereby we suppose that things which can never be seen or touched and only exist in our imagination and understanding can have the properties and relations only of things which can be touched. We impose on the creations of thought the limitations of the visible and tangible.[3]

Mach was not wrong in this. Atoms do *not* respect such limitations; the laws of atomic physics (which did not begin to emerge clearly until 1925) are not those of the "visible and tangible." When Einstein and Mach met late in 1913 and discussed these matters, Mach agreed that atoms were a useful physical hypothesis, but he did not concede the "real existence" of atoms.

"Mach recognized the weak spots of classical mechanics," wrote Einstein. What did he mean by that? Newton's principle of inertia is expressed in his first law of motion—that a body not acted on by a force continues in its state of uniform motion or rest. That statement takes for granted a frame of reference, which Newton identified as "absolute space." Mach objected to "absolute space" as being an extraneous element, writing, "For me only relative motion exists." He considered the reference frame to be supplied by the "fixed stars" and was thereby led to the concept that the inertia of any body is ultimately attributable to the *physical* influence of the distant *matter* of the universe. To Einstein, Mach's dethronement of "absolute space," a space that controlled physical phenomena but was not influenced by those phenomena, implied that the geometry of space (-time) had to be part of the physical mechanism of gravitation.

It is important to understand why Mach himself could make no progress with his ideas on the basis of the then-existing theory of gravitation. We know that different inertial observers are in uniform relative motion. And it is a matter of experience that inertial reference frames are those in which the "fixed stars" do *not* rotate (this is the precise sense in which the "fixed stars" supply the reference frame). As mentioned in Chapter 1, an observer in *rotation* relative to an inertial observer is *not* an inertial observer. What is the fundamental difference between a change of reference frame involving a small straight-line or translational motion and one involving a small rotational motion? Just this. After a small translational change, the distant matter of the universe is perceived to have a *small* translational motion, in the opposite sense. After a small rotational change, the distant matter of the universe is perceived as moving (to the extent that Euclidean geometry applies) with a speed that increases proportionally to the distance of the matter; that speed can become very *large*. Here is indeed a fundamental difference, but Newton's law of gravitation gives no *dynamical* expression of it; Newtonian gravitational forces depend on position, not on motion. And that is why the general theory of relativity, with its motional effects, creates a wholly new situation.

Earlier in the year that witnessed their meeting, Einstein had written to Mach about relativity and gravitation.[4] He first spoke of a solar eclipse "next year" (1914) that would show whether light rays were bent by the Sun.* He then went on to write (remember that this was two years before the final theory was produced):

> The following additional points emerge:
>
> 1. If one accelerates a heavy shell of matter, then a mass enclosed by that shell experiences an accelerative force.
>
> 2. If one rotates the shell relative to the fixed stars about an axis going through its center, a [motional] force arises in the interior of the shell; that is, the plane of a Foucault pendulum is dragged around.

What is a Foucault pendulum?

Foucault's Pendulum

Jean Foucault, foreign member of the Royal Society, measurer of the speed of light, inventor of a gyroscope in 1852, had, the preceding year, suspended a 60-m-long pendulum from the dome of the Panthéon in Paris. He set it in motion by igniting a cord that held the heavy bob to one side. A ring of sand on the floor was cut, on both sides of the swing, by a pointed rod mounted under the

*In that year, some German astronomers went to a certain region along the path of totality. It was the wrong place, at the wrong time. They were interned in Russia for the duration of the war.

Foucault's pendulum (left); a modern Foucault pendulum (right)

bob. From the changing position of the notches in the sand it became apparent that the plane of the swing was not holding steady: it was rotating slowly in the *clockwise* direction. Yet nothing in the arrangement seemed to prefer one sense of rotation to the other.

What it is that needs understanding becomes simpler by performing this experiment *al fresco,* with the pendulum supported over Earth's North Pole. The plane in which the pendulum swings turns clockwise over the ground. Looking up at Polaris directly overhead (this experiment can be done *anytime* in the winter), we see the surrounding stars also majestically turning, in step with the pendulum! The pendulum is locked on the stars, while the Earth turns *counterclockwise,* underneath. We traveled to the North Pole because only there (in the northern hemisphere) does the suspension point of the pendulum remain fixed relative to the distant stars. Elsewhere, the suspension point rotates with the Earth about the polar axis and this produces a counter rotation,* which diminishes the effect. The two motions balance at the equator, where a Foucault pendulum does *not* rotate relative to the ground. The period of rotation of the plane of the pendulum's swing is about 32 h at Paris (latitude 49° N); at the North Pole it is, of course, 23 h, 56 m: one sidereal day.

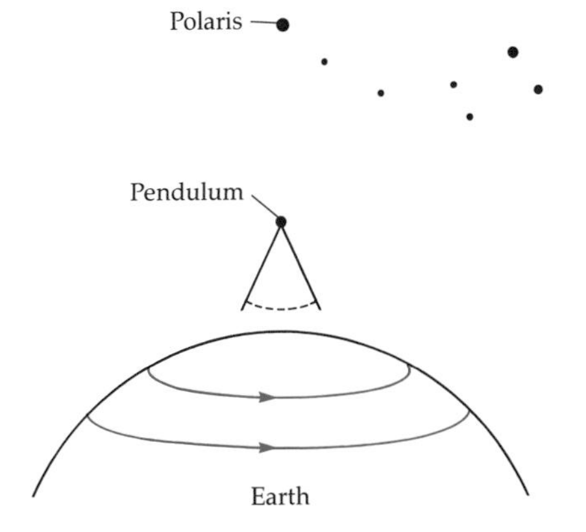

In 1913, Einstein already knew that, if a Foucault pendulum, with its plane of oscillation locked on the stars, were surrounded

*To see this, suspend a weight on a string and give it a pendulum motion, using your hand as support point. Now move your hand slowly in a horizontal circle.

by a spinning shell of matter, it would acquire a *tiny* fraction of that rotary motion: "the plane of the Foucault pendulum is dragged around." Here was confirmation of Mach's hypothesis concerning the influence of distant matter. It became plausible that the preponderant mass of the universe, in whatever state of rotational motion it might be, would impart its rotary motion *fully* to all other bits of matter, thereby producing the natural *absence* of rotation relative to the distant stars.

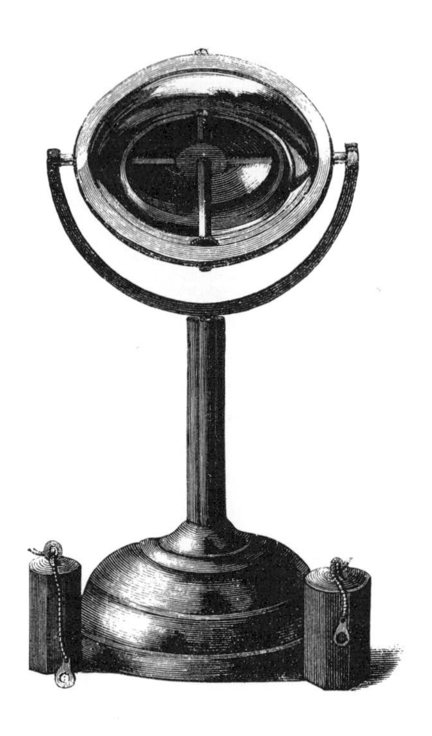

Foucault's gyroscope

Gyroscopes

So it seems that the stars do influence our lives. It would be nice, however, if we could bolster this awesome vision, of gravitation acting across the whole universe to make it spin as one, by some direct experimental evidence a little closer to home. Home is the Earth, and it rotates. Presumably, the rotating Earth does drag the plane of pendulum motion with it. Unfortunately, the estimated time for the Earth to drag that plane full circle is measured in millions of years. To detect any such motion requires a long observation time, as well as a pendulum that keeps beating steadily despite air resistance and friction. It is clear that the experiment needs an air-free environment, a vacuum. That suggests an Earth-satellite experiment. But, within the satellite, which is falling freely in Earth orbit, there is *no* gravity; a pendulum will not work. Foucault would have known the answer to that difficulty; a rapidly spinning gyroscope also locks on the stars. Gyroscopes are now in everyday use as a replacement for the stars in pinpoint navigation under the sea and in the air.

A satellite-gyroscope experiment has long been in the planning stage at Stanford University. Before examining some of the technical wonders of this experiment, it will help to understand the gravitational motional effects that it is designed to detect through an analogy with a related area of experience in which motional forces are more familiar: electromagnetism.

BOX 6·1 ***The First Electric Telegraph***

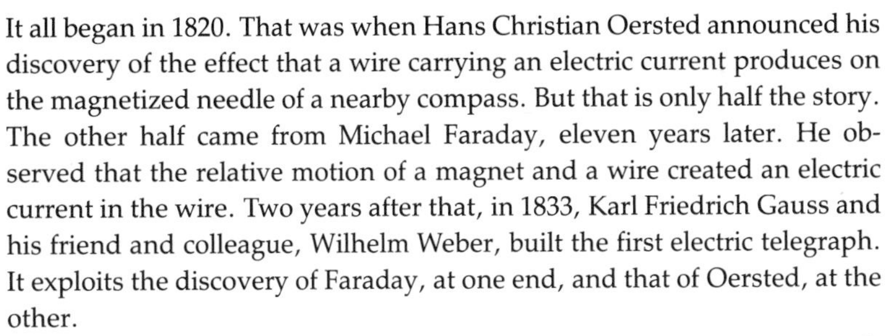

It all began in 1820. That was when Hans Christian Oersted announced his discovery of the effect that a wire carrying an electric current produces on the magnetized needle of a nearby compass. But that is only half the story. The other half came from Michael Faraday, eleven years later. He observed that the relative motion of a magnet and a wire created an electric current in the wire. Two years after that, in 1833, Karl Friedrich Gauss and his friend and colleague, Wilhelm Weber, built the first electric telegraph. It exploits the discovery of Faraday, at one end, and that of Oersted, at the other.

The rapid up-and-down motion of a magnet placed between wire coils produces two short pulses of electric current (Faraday). That was the basis of the transmitter. The detector was a coil above which was suspended a magnet that, by itself, maintained a northerly direction. When the current pulses passed through the coil, the magnet was deflected (Oersted), which could be observed by the effect on a light beam reflected from a mirror mounted on the magnet. Sender and receiver were connected by a pair of wires that formed a closed circuit. By reversing the connections at the transmitting end, two kinds of signals could be sent, resulting in a deflection of the detecting magnet to the left or to the right. A code was devised for transmitting the letters of the alphabet. All of eight letters a minute could be sent.

With considerable difficulty, Weber strung iron wires between Gauss's observatory, outside the walls of Göttingen, over the north tower of St. John's church, to Weber's laboratory on the banks of the canal that carried the Leine River through the town; a distance of several kilometers. The story goes that the very first message from Weber to Gauss announced that an assistant was coming from Weber's laboratory to the observatory. By the time the message was decoded, the traveler had arrived. Nevertheless, Gauss was fully aware of the awesome implications of such a communication device, "before which the imagination almost reels." But it was not possible for him to participate in its practical development. His limited budget (150 talers a year) was fully committed to his astronomical and Earth magnetism observatories. Enter Samuel Morse.

Yet Gauss did stimulate a certain Steinheil to further develop this telegraphic technique, and in 1838 he confirmed Gauss's important suggestion that a closed wire circuit was unnecessary; the conducting Earth could serve as the return path. The end of the first electric telegraph came before Christmas of 1845 when a lightning bolt struck the church tower and destroyed the wire. Thirty years later, there were 175,000 km of telegraph wire in the United Kingdom alone.

ELECTROMAGNETISM AND GRAVITATION

Similarities and differences between electromagnetism and gravitation are already evident in what we know about slowly moving bodies. Both kinds of forces vary inversely as the square of the distance between any two small bodies. The electric force, proportional to the product of the charges, can be a force of repulsion or of attraction, according as the charges are like (same sign) or unlike (opposite sign). But the gravitational force, proportional to the product of the masses, is *always* one of attraction. More impressive is the enormous disparity in the strengths of the forces. As an example, consider two equally charged atomic particles—two protons, say, at a given distance from each other. To duplicate the strength of that (repulsive) electric force in the (attractive) gravitational force between two equally massive bodies, at the given distance, each body would have to be composed of 10^{18} protons.*

Rotation

We are interested in the gravitational effects of massive rotating bodies. What happens *electromagnetically* when a sphere of positive electric charge, for example, spins about an axis? The rotational motion produces a *magnetic* field. Outside the rotating sphere, the magnetic field is just like that of a small bar magnet with its north-

*If you can't wait to rush out and check this experimentally, be sure *not* to use naked protons; use hydrogen (proton *and* electron) atoms, which are electrically neutral. Only by utterly quenching the vastly stronger electric forces can the force of gravitation be sensed at all. Besides, how would you hold all those protons together?

south line pointing along the axis of rotation. We can prove that just as Faraday would have, by sprinkling iron filings on a sheet of paper held over the rotating sphere. The patterns formed by the filings disclose the magnetic lines of force in space; these patterns are similar to the lines of force of a bar magnet. The strength of this magnetic field varies as the inverse *cube* of the distance. It also depends on direction; if we compare points at the same distance, it is twice as strong in a line with the poles than at right angles to the line of the poles, for example.

As far as the outside magnetic field is concerned, the sphere of positive charge could be a thin *shell* of charge. Then, what happens inside the spinning shell? There is a magnetic field inside, too; it is a *uniform* field, with the same direction (that of the spin axis) and the same strength everywhere within the spherical shell.*

How does a moving, negatively charged particle behave inside the spinning positively charged shell? The magnetic field gives the charged particle a rotary motion, precisely the kind of circular motion that takes place in particle accelerators (see Chapter 3). For a negative charge inside a spinning distribution of positive charge, the sense of that imparted rotation is the same as the sense of the

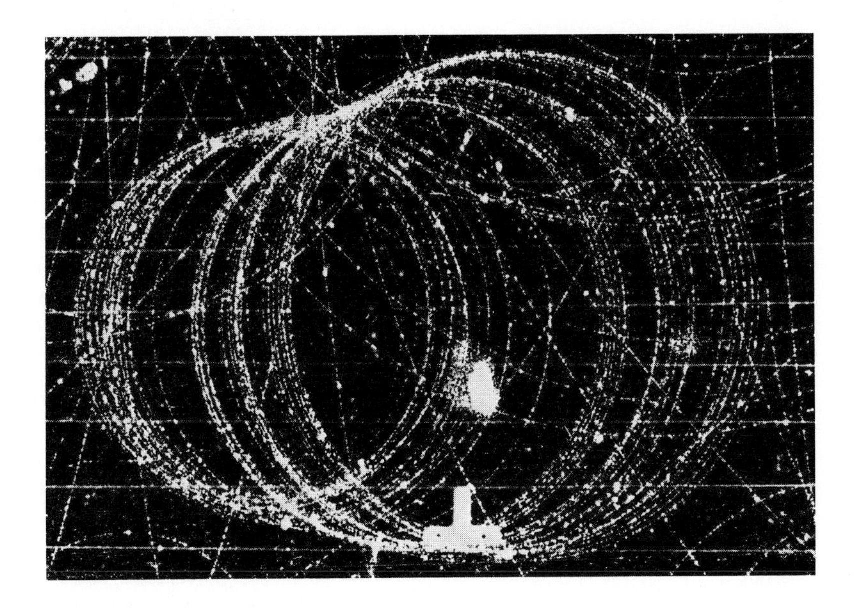

The circular path of an electron in a magnetic field

The diameter of the circle changes as the electron moves into a region of weaker field.

*The electric field within the shell is also uniform—uniformly *zero*. That is the unique property of the inverse square law of electric force.

spin. The rate of rotation is related to the acceleration of the particle, which is given by the ratio between the force on it (proportional to its electric charge) and its inertial mass. This ratio of charge to mass varies from one kind of particle to another (it is 1,836 times as large for an electron as for a proton).

Now let us work by analogy and replace electromagnetism by gravitation, which our choice of *opposite,* attracting charges anticipates. Giving a spin to a massive spherical shell should impart a rotation, in the same sense as the spin, to any particle in the interior of the shell. And the charge-to-mass ratio of electromagnetism becomes the ratio of *gravitational* mass to *inertial* mass, which is a universal constant. The spinning massive shell drags matter with it, imparting to matter of every kind a common rotation. The analogy works.

SATELLITE GYROSCOPE EXPERIMENT

In describing this experiment, when a certain body is called a "magnet," it means that the body has gravitational properties analogous to the electromagnetic properties of a real magnet. Thus, the massive spinning Earth is a "magnet," as is the gyroscope in the orbiting satellite. Owing to the angle dependence of the magnetic field of a bar magnet, the two "magnets" exert twisting forces on each other. In particular, if the gyroscope is located over the Earth's North Pole, with the gyroscope spin axis perpendicular to Earth's polar axis, the axis of the gyroscope should rotate about the polar axis. *That* is a deviation from Newton's laws, according to which the gyroscope should hold steady on the framework of the distant stars.

Of course, we cannot hang the gyroscope in space over the North Pole. The nearest equivalent is to put the satellite carrying the gyroscope into an orbit that passes over both poles. And that introduces something else, associated with the motion of the gyroscope. Relative to the "magnet" that is the gyroscope, the massive Earth is a moving "charge," an "electric" current, producing a "magnetic" field. That "magnetic" field also acts to twist the gyroscope. This effect is greatest if the gyroscope spin axis is in the plane of the orbital motion (it changes direction in that plane) and disappears if the spin axis is perpendicular to the orbital plane.

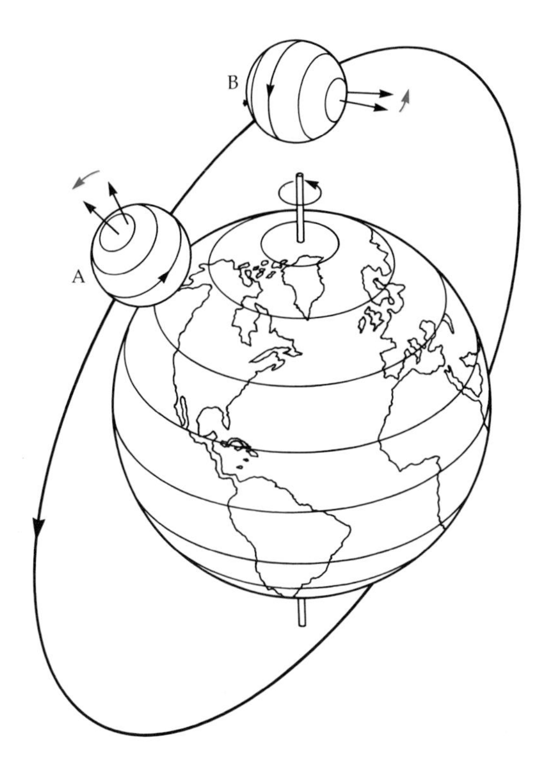

Satellite gyroscopes

The spin axis of gyroscope A is in the orbital plane. The spin axis of gyroscope B is perpendicular to that plane.

Because we are free to select the spatial orientation of the gyroscope, we can separate the effect of the orbital motion of the satellite from the effect of the spin of the Earth.

How big is the expected orbital effect? Through what fraction of a circle does the gyroscope spin axis move in the orbital plane, during one circuit of the Earth? The answer to this question, as to several others—gravitational red shift, light deflection, time delay—is proportional to the strength of the gravitational field, in the combination $g'R/c^2$, where g' is the acceleration of gravity at R, here the radius of the orbit. For an orbit that grazes (airless) Earth, where the acceleration is g ($10\ \text{m/s}^2 = 10^{-2}\ \text{km/s}^2$), at the distance of the Earth radius R (6.4×10^3 km), the quantity gR/c^2 is

$$\frac{gR}{c^2} = \frac{(10^{-2}\ \text{km/s}^2)(6.4 \times 10^3\ \text{km})}{(3 \times 10^5\ \text{km/s})^2} = 7 \times 10^{-10}.$$

Only the additional factor 3/2 is needed to produce the prediction of general relativity: the gyroscope with its axis in the orbital plane should turn through 1.05×10^{-9} of a circle in one revolution about the Earth.

A practical experiment would have to accumulate this effect over many orbits. How many are there in a year for an Earth-grazing satellite? Since the time of Sputnik (October 4, 1957), it has been familiar that a satellite in a low orbit revolves about the Earth in about 90 minutes. That period would be 84 minutes for the hypothetical Earth-grazing satellite.* On dividing this period into the

---

*This is hardly an isolated fact. The reader is invited to start with the following properties of the Moon, revolving about the Earth—sidereal period, 27.32 days; radius of orbit, 3.844×10^5 km—and apply Kepler's third law of planetary motion, which says that the square of the orbital period is proportional to the cube of the radius of the orbit. What is the period for a revolution about the Earth in an orbit of radius 6.378×10^3 km?

Here's another way. Moving at speed v in its circular orbit of radius equal to Earth's radius R, the Earth-grazing satellite is held in orbit by the inward acceleration (Box 4.2) $v^2/R = g$. The satellite speed, $v = \sqrt{gR}$, multiplied by T, the period of the orbit, equals the circumference of the Earth, $2\pi R$. Thus, the period is

$$T = 2\pi\sqrt{R/g},$$

which also happens to be the period of a pendulum of length R, swinging in a *uniform* gravitational field of acceleration g.

The equivalence of the two calculations is a replay of Newton's discovery (ca.1670) that gravity on the surface of the Earth, extended to the Moon by the inverse-square law, would keep the Moon in its orbit.

number of minutes in a year (5.3×10^5), we get the number of revolutions in a year: 6.3×10^3. So the fraction of a circle through which the gyroscope axis turns in a year is expected to be (6.3×10^3) (1.05×10^{-9}) = 6.6×10^{-6}. We convert this into an angle by noting that a full circle is $360 \times 3600 = 1.3 \times 10^6$ arc sec, which gives somewhat less than 9 arc sec/yr. In a more realistic orbit at an altitude of one-tenth of the Earth's radius, this is reduced by about 25 percent to 7 arc sec/yr; that yearly change in direction is, for example, the angle subtended by 13 km at the distance of the Moon.

When the gyroscope axis is perpendicular to the satellite orbital plane, this motional effect is removed, and only the rotation of the Earth acts to change the direction of the spin axis of the gyroscope. A point on Earth's equator moves more slowly than the grazing satellite does, seventeen times as slowly. The preponderance of Earth's matter is nearer the center, however, and has a significantly smaller speed than the equatorial point. That, combined with the variation of the spinning Earth's "magnetic" field over the satellite's path, reduces the effect considerably. For a satellite in a polar orbit, at an altitude of one-tenth of the Earth's radius, the axis of the gyroscope should be dragged in the same sense as the Earth's spin by 0.05 arc sec/yr; the yearly amount is the angle subtended by 100 m at the distance of the Moon.

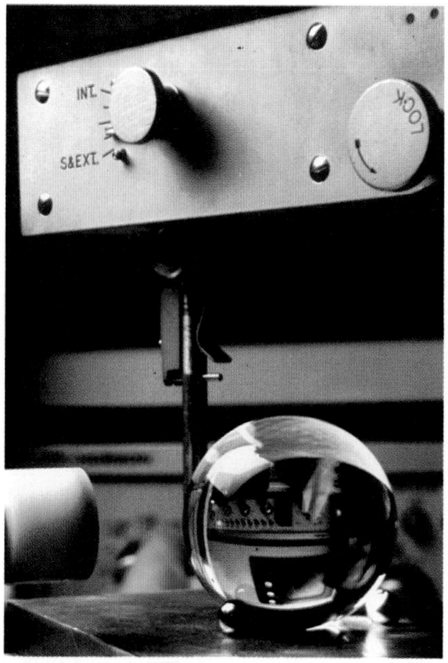

The quartz ball of the frictionless gyroscope

Superconductivity

By what miracles of technology do we expect to measure such tiny angles? How could one produce a frictionless gyroscope that—apart from these sought-for effects of general relativity—would hold its fix on the stars to better than a hundredth of a second of arc per year? Hope rides on the technology of extremely low temperatures, just a few degrees above absolute zero, which is the temperature of totally dark and empty space. The decisive phenomenon is *superconductivity:* some substances, at sufficiently low temperatures, conduct electric currents with absolutely *no* resistance, *no friction*.

First, we need a spinning body to serve as the gyroscope. To avoid having this body twist from other causes than the one sought for, it must be almost perfectly uniform. A quartz ball, 4 cm in diameter, with a density that is homogeneous to better than one part in a million, and spherical to several parts in ten million (its biggest bump is proportionally no higher than a man standing on a spherical Earth), is coated with a film of the superconducting metallic element niobium.* The ball is then suspended in the weightless environment by applying small voltages to three mutually perpendicular pairs of electrodes. Next, the ball is spun up to 200 revolutions a second by jets of gas; the gas is then pumped out, leaving the ball running freely in a nearly perfect vacuum. This is the *frictionless gyroscope.*

Very good, but useless, unless one can sense with great accuracy the direction in space of the gyroscope spin axis. Here, again, superconductivity works for us. In a superconductor, the flow of negatively charged electrons is indifferent to the positively charged atomic cores. Conversely, the motion of the atomic cores produced by the spin of the ball does not influence those electrons. The resulting relative motion of positive and negative charges is an electric current, the source of a magnetic field; a spinning superconductor acts like a weak magnet with its poles along the axis of spin. Any change in this magnetic field, as the spin axis moves, can be detected through the currents induced in surrounding superconducting loops.

These minute currents are sensed and measured by a SQUID—not a creature of the deep, but the acronym of a remarkable instrument, a S(uperconducting) Q(uantum) I(nterference) D(evice). It relies on the fact that the number of magnetic lines of force threading a superconducting loop can change by only a discrete amount, a quantum jump. But how can one expect to measure such tiny changes in the magnetic field, in the presence of much larger fields of uninteresting origin? Superconductivity again. A superconductor acts as a nearly perfect shield against magnetic fields. If the working parts of the gyroscope are surrounded with such a shield, any tiny residual field that remains will stay very constant, permitting the small changes of interest to be detected.

Given a frictionless gyroscope, and the ability to sense its motion, we must relate that motion to the framework of the distant stars. A telescope is needed, with its line of sight held fixed on some convenient star, such as Rigel in the constellation Orion. All this is put into a spacecraft, intended to be launched into orbit from a space shuttle. Now we come to some space wizardry.

The theory that we want to test assumes that the gyroscope moves only under the influence of Earth's gravitational field. But

*This element was first called columbium, after the United States; a mineral specimen from Massachusetts that was in the British Museum collection led to its identification in 1801.

that gyroscope is housed in a satellite, which feels the slight drag of the residual atmosphere, present even at that altitude, and experiences the pressure of solar radiation. To compensate these effects and produce a drag-free satellite, another ball is suspended in the interior of the satellite, to act as the ideal particle of the theory. The ball's supporting system senses any displacement of the satellite relative to the ball and activates thrust jets on the satellite that compensate the displacement. The gas for these and other control jets is helium, boiled off the liquid helium that maintains the low temperature of the superconducting devices.

Truly the gyroscope satellite experiment is at the frontier—both scientifically and technologically. Sometime in the 1980s, after being carried up from Earth by a space shuttle, this remarkable apparatus will begin its lengthy task.

BOX 6·2 ***Gravitational Images***

Einstein, in 1936, pointed out that the bending of light rays near a massive body might have a bizarre astronomical consequence. A star seen near the Sun's edge has its light bent so that it appears to be displaced *outward*. Now, suppose that, in the line of sight between Earth and a very distant astronomical object (a quasar, for example), a very massive and compact body intervenes. Light passing on one side of that body is bent in such a way that the quasar image is displaced outward on that side; light passing on the other side of the body is bent in the opposite sense and produces another outward displaced image. One would see two images of the same quasar.

In 1979, it was recognized that the radio source named 0957 + 561 is associated with a pair of quasars that are separated in the sky by only 6 seconds of arc. These quasars have remarkably similar optical properties. Here, it seemed, was a realization of Einstein's prediction. Subsequent radio investigations were consistent with this interpretation but disclosed additional features that did *not* fit with the simplest model of the intervening object: a compact massive body. Then in 1980 and 1981, optical studies turned up a faint elliptical galaxy, with a related cluster of galaxies, only 1 arc sec away from one of the images. The distribution in space of these galactic bodies *is* compatible with the features observed at the various wavelengths.

Another triumph for Einstein.

GRAVITATIONAL WAVES

The gravitational forces of Newton, which act between masses, are analogous to the electric forces between charges. The satellite gyroscope experiment is designed to test whether there are also gravitational forces analogous to magnetism. In the hands of Maxwell, the union of electricity and magnetism led to the prediction of electromagnetic waves. We live in a world where a whole spectrum of such waves, from radio waves to X-rays, is commonplace. Well, then, are there also gravitational waves? Einstein predicted them immediately upon creating the general theory of relativity. And he said that gravitational waves travel at the same speed as electromagnetic waves, at speed c.

Underlying this expectation of equality between the speeds of the two kinds of waves is the similarity in the two kinds of forces between slowly moving bodies at sufficiently large distances: both vary inversely as the square of that distance. One way to see this begins with Einstein's 1905 discovery that light has both wave and particle properties. The subsequent development of quantum mechanics, the laws of atomic physics, made it clear that this dualism between particle and wave (or field) is universal. Indeed, the fact that particles such as electrons, protons, and neutrons also have wave properties is fundamental to our understanding and control of atomic and nuclear behavior. And so, if there are gravitational waves, there must also be gravitational particles, *gravitons.* What was first learned for the photon, that the particle *energy* is proportional to the frequency of the wave, or *inversely* proportional to its *period,* holds generally for particle waves. The *momentum* of a particle, analogously, is, with the same constant, *inversely* proportional to the *wavelength.* Thus, for light, the particle relation $E = pc$, or $E/p = c$, is also the wave relation equating the speed c to the ratio of the wavelength and the period (the distance traveled in one period is the wavelength).

The question of the speed of gravitational waves now takes this form: does the graviton have zero rest mass, like the photon, or does it have a nonzero rest mass? The picture that quantum mechanics supplies for the electric and Newtonian forces between slowly moving, distant bodies is that the modest amounts of energy and momentum exchanged between such bodies are carried by a photon and a graviton, respectively. On the atomic scale, the energy (E) and momentum (p) associated with these particles are very small—essentially *zero*. That is indeed permissible for a zero-rest-mass particle, for which the values of those mechanical properties (see Box 3.3) are related by $E^2 - (cp)^2 = 0$. But, for a particle with a rest mass m_0, for which $E^2 - (cp)^2 = (m_0c^2)^2$, it is impossible. That is why both photon and graviton are particles without rest mass, traveling at the speed c.

Although experimental evidence for the wave nature of light goes back to the beginning of the nineteenth century, it was not until 1923 that clear evidence for the particle aspect of light was found (see Chapter 3). We can expect the same order of discovery for the gravitational analog of light; the time scale of these discoveries is another matter. The quest for gravitational waves is at the frontier of research and is being pursued at a number of laboratories around the world.

Difficulties

If gravitational waves exist, why should it be so difficult to detect them? One glaring reason has already been mentioned: under ordinary circumstances, gravitational forces are vastly weaker than electromagnetic forces. It is only for bodies on an astronomical scale that gravitation becomes a dominant force. It will be useful, then, to introduce new measures on that scale that are gravitational in nature.

The Newtonian force varies inversely as the square of the distance and is proportional to the product of masses. Let us call the constant of proportionality G. What is its value? That depends on the choice of units for mass, length, and time. As discussed in Box 6.3, the natural value of G for describing gravitational phenomena, $G = 1$, can be produced by adopting units based on Earth's Newtonian motion about the Sun. The unit of mass is the solar mass, the unit of length is half the largest diameter of Earth's elliptical orbit, about 150 million km, and the unit of time (in a simplified form applicable to a circular orbit) is that required for the Earth to move through an angle of one radian: the sideral year divided by 2π, about 58 days. In contrast with ordinary units—gram, centimeter, and second, for example—these units are huge (in varying degrees).

Instead of the natural, Newtonian units for gravitation, one can introduce natural Einsteinian units. Keeping the solar unit of mass, the unit of length is shortened to about 1.5 km, and the unit of time becomes nearly 5 μs. That retains $G = 1$ and achieves $c = 1$.

BOX 6·3 *Gravitational Units*

For simplicity, think of Earth's orbit as circular, which it nearly is. Earth maintains its position, at the distance R from the Sun, by a balance between the inward acceleration produced by the gravitational force, GM/R^2, in which M is the solar mass, and the outward acceleration of rotational motion, v^2/R (see Chapter 4). Here, v, the speed in the orbit, is also the ratio of the circumference of the orbit ($2\pi R$) to the time required to traverse that distance (T, the period of revolution, which is one year). This balance gives us the explicit form of Kepler's third law, as produced by Newton:

$$(2\pi R/T)^2/R = GM/R^2,$$

or

$$(2\pi/T)^2R^3 = GM.$$

The universal gravitational constant G is related to the measures of mass, length, and time. What if we were to adopt the Sun's mass as our mass unit? Then $M = 1$. How about the radius of Earth's orbit as the unit of distance? Then $R = 1$. Now that we are in the spirit of this, let us take $T/2\pi$, the time required to go through an angle of one radian (57.3°) around the orbit, as the unit of time. Then $T/2\pi = 1$. In these *Newtonian* gravitational units, $G = 1$.

As noted many times, the ratio of the Earth's speed in its orbit to the speed of light, v/c, is very close to 10^{-4}. In Newtonian gravitational units, $v = 2\pi R/T$ is unity, which gives c a value that is nearly 10^4. Can we find other gravitational units that retain $G = 1$, and also make $c = 1$? Yes. Let us keep the *solar* unit of mass but change the units of length and time in a way that maintains the ratio between the cube of the length unit and the square of the time unit (Kepler's third law). Then $G = 1$ persists. That is done when we *shorten* the length unit by a factor of 10^8 and the time unit by a factor of 10^{12}: $(10^8)^3 = (10^{12})^2$. With a unit of speed that is $10^{-8}/10^{-12} = 10^4$ times as large, $c = 1$. What are these *Einsteinian* gravitational units, in which $G = 1$ and $c = 1$? The unit of distance, 10^8 times shorter than the radius of Earth's orbit (which is 1.5×10^8 km), is very nearly 1.5 km. An earlier discussion of units for which $c = 1$ presented the example of 0.3 km and 1 μs (Chapter 5). The unit of distance now being advocated is five times as large and therefore so is the time unit; the Einsteinian gravitational unit of time is 5 μs.

It is worth noting that $g'R/c^2$—or, as we are now able to write it ($g' = GM/R^2$), GM/c^2R—has the value unity when both R and M are one Einsteinian unit.

More subtle than the disparity in strength between gravitation and electromagnetism is the different character of the respective *charges*. Electric charge is either positive or negative; gravitational charge, mass, is always positive. We can appreciate the consequences of that by thinking of a simple model of a wave generator: two charges of equal *magnitude* at the ends of a vibrating spring. A *positive* electric charge moving in one direction is an electric current. An equal electric current is produced by a *negative* charge moving in the *opposite* direction. The two currents reinforce each other. But, if the two charges have the *same* sign, they cancel each other.

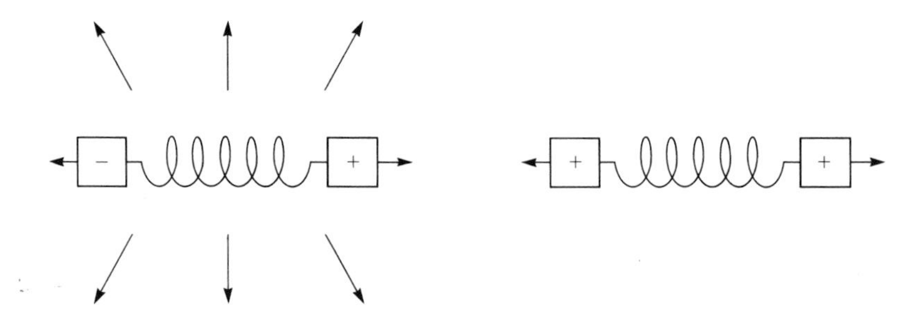

This is just the situation for a *gravitational* wave generator using masses of equal magnitude. Perhaps you think: make the masses unequal. Ah, but Newtonian mechanics steps in to say that the more massive body will move proportionately more slowly, and the cancellation remains (more about that later).

Does this mean that there is no gravitational radiation, after all? No, but it does supply another reason why it is difficult to emit and detect gravitational radiation. What has just been said about wave generators is correct only if the wavelength of the radiation is *enormous* compared with the separation between the two bodies. When this is not the situation, the waves emitted by the two accelerated bodies travel significantly *different* distances to the detector (in units of the wavelength), and the cancellation is not complete. The fraction that survives is roughly measured by the average sep-

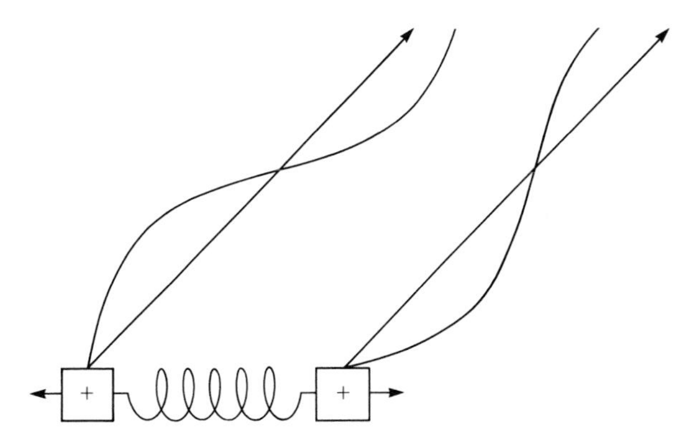

Incomplete cancellation

Here the separation between charges is not small compared with the wavelength.

aration of the two bodies divided by the wavelength. The reciprocal of the wavelength is also equal to the frequency divided by c. The product of the frequency and the distance between the bodies is a measure of the rate at which that distance changes in time, which is a speed v. So the multiplicative factor expressing the incomplete cancellation of the waves is v/c.

Waves have amplitudes; as mentioned earlier, the amplitude measures the height of the crest of the depth of the trough. Waves carry energy, which is determined by the *square* of the amplitude. The gravitational wave emitted by *one* of the two oscillating bodies has an amplitude proportional to its mass and to the *acceleration* of the body. (An unaccelerated mass, one with uniform velocity, does not radiate. To an observer moving with it, the velocity of the body is zero; it has no kinetic energy to emit.) Notice that the product of mass and acceleration is the *force* acting on this body, which is just the negative of the force acting on the other body. That makes quite explicit that the cancellation for very long wavelengths is independent of the choice of masses. The rate at which energy is radiated by the *two* oscillating masses is therefore given, apart from constant factors, by the product of the square of the force with the square of the relative speed of the two masses.

Now let us use the spring with attached charges as a model of a wave *detector*. An electromagnetic wave falling on a nonvibrating spring with unlike charges will push those charges in opposite directions (as shown in the left-hand diagram at the top of the next page) and set the spring vibrating. That effect is particularly pronounced when the frequency of the wave matches the vibrational

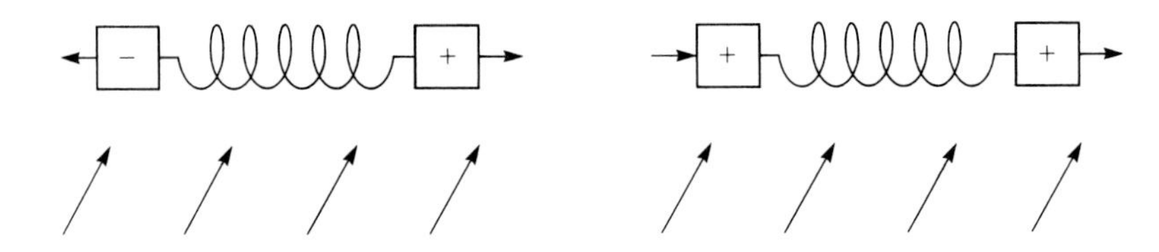

frequency of the spring—when they are in tune. If the charges are the same, however, they are pushed in the same direction, and the spring is not affected. But this device can still detect *electromagnetic* waves because one can observe the motion induced in the whole system. Well then. Can one observe gravitational waves in the same way, by the common motion induced in the masses and the spring? No!

For electromagnetism, the *observing* apparatus is electrically neutral—its natural state—and is unaffected by the electric force pushing the detector. But with gravitation, the observing apparatus (which includes the observer) has a mass and is affected by the gravitational wave in precisely the same way as the detector, and so *no* relative motion will be discernible. It is the old story of the uniform gravitational field. Again, this ceases to be entirely true in proportion to the ratio of the distance between the masses and the wavelength. Because the field is *not* uniform (it varies spatially on the scale set by the wavelength), the two masses experience somewhat different accelerations, and the spring *is* set into vibration.

Sources

Gravitational waves are hard to generate and hard to detect—the opposite sides of the same coin. To succeed in proving the reality of gravitational waves, one must begin by identifying the strongest conceivable sources of such waves. As for possible gravitational wave emitters, how about a planet swinging around a star? Earth itself, for example. As indicated in Box 6.4, the Earth does radiate gravitational waves, with about the power of an ordinary desk lamp. Pretty poor performance. We can understand the reason for it by looking at the numbers that characterize the Earth in its orbit about the Sun, as they appear in Einsteinian gravitational units (solar mass; 1.5 km; 5 μs).

In these units, the mass of the Earth is 3×10^{-6}, the distance of the Earth from the Sun is 10^{8}, and the speed of the Earth is 10^{-4}. The gravitational force exerted on the Earth is the planet's mass divided by the square of the distance: 3×10^{-22}. The square of that

BOX 6·4 ***Earth's Gravitational Glow***

The rate at which an accelerated body radiates energy in the form of gravitational waves is measured by the product of the *square* of the *force* acting on it and the *square* of its *speed* relative to c. Let us use Einsteinian gravitational units, in which $G = 1$, $c = 1$, and the Sun sets the standard of mass. On that scale, the Earth's mass is 3×10^{-6}. The force between Earth and Sun is measured by the mass of the Earth divided by the square of their distance, which distance is 10^8 in these units; Earth's speed is 10^{-4}. So far we have

$$(3 \times 10^{-6}/10^{16})^2 \times (10^{-4})^2 = 9 \times 10^{-52}.$$

All this is multiplied by the *Einsteinian unit of power:* one solar-mass equivalent of energy emitted in 5 μs. The energy in one solar mass (2×10^{30} kg) is prodigious, 1.8×10^{47} J(oules), and this Einsteinian unit of gravitational power is a mind-boggling 3.6×10^{52}W(atts) when expressed in Earth-bound units of power: 1W = 1J/s. But, when the gigantic gravitational power unit is multiplied by the Earth-related factor, it becomes a mere $3.6 \times 9 = 32$ W. This needs only the numerical factor 32/5 to be the actual general relativistic prediction of the power with which the Earth shines gravitationally—200 W, the output of a decent desk lamp.

force multiplied by the square of the Earth's speed gives the Earth's rate of emission of gravitational energy: 9×10^{-52}, which is TINY. It becomes clear that a powerful source of gravitational radiation requires masses, distances, and speeds that are comparable to the Einsteinian units—solar masses, kilometer distances, and speeds approaching that of light.

Kilometer distances? That tells us where to look: stellar catastrophes. A massive star that has used up its store of nuclear fuel suddenly collapses under its own gravitational forces to form a neutron star or possibly a black hole, with dimensions that are measured in kilometers. Now what can we expect for the rate of emission of gravitational waves, the power of such a source? It would be some significant fraction of one Einsteinian unit of power: one solar mass of energy emitted in 5 μs, or 200,000 solar-mass equivalents per second!

Crab Nebula

The sudden flare-up of a supernova* signals a stellar collapse. The supernova of July 1054 is recorded in the annals of the Chinese and possibly in petroglyphs of the American Indians. The neutron star that resulted is now observed as a pulsar, flashing out beams of light and radio waves thirty times a second, at the center of the Crab Nebula, the still-expanding debris of that explosion.

Suppose that a massive star collapses to form a black hole at the center of our Galaxy, 30,000 light years distant. It is thought that the last stage of collapse takes about a thousandth of a second, during which the star radiates gravitational waves with frequencies of about 1,000 Hz and wavelengths of hundreds of kilometers. If the energy of one solar mass is emitted uniformly in all directions, the energy that reaches the Earth is about what the Sun radiates electromagnetically into our atmosphere in one hundred seconds.

---

*It is estimated that a supernova appears about every twenty-five years in our Galaxy, but owing to the obscuring dust and gas within the Galaxy (we sit out toward the edge) only five supernovae have actually been seen within the Galaxy in the past two thousand years.

That is a *prodigious* amount of energy, but, in the form of gravitational waves, it is still fantastically difficult to detect in a terrestrial laboratory.

Detectors

The early 1960s witnessed the first acceptance of this formidable experimental challenge. The detector that was then developed was, in essence, the masses and spring discussed earlier. It took the form of a large aluminum cylinder. The two halves of the cylinder play the role of the masses. The analog of the spring is the slowest vibration of the cylinder, the deepest tone it sounds when struck at its ends like a two-headed drum; it rang at 1660 Hz, nearly two octaves above the standard A tone. The bar was suspended mechanically in a vacuum chamber and insulated from the noise of the outside world. Attached to the bar were crystals that converted the motion to be detected into electric currents. (Inside a quartz watch is just such a crystal acting as a very stable oscillator.)

Just how much of a drum tap would there be in our example of a gravitational pulse coming from the center of the Galaxy? The

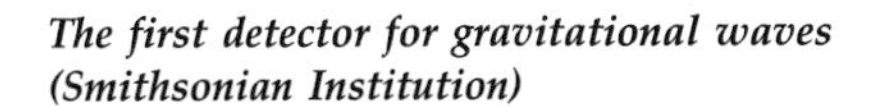
The first detector for gravitational waves (Smithsonian Institution)

ends of the bar would move by all of 10^{-15} cm, one-ten millionth of the diameter of an atom. Very delicate drum playing. That distance is impressively small, but the crucial experimental difficulty becomes clearer if we talk in terms of *energy*. The bar is there as a practical way of supplying masses for the gravitational wave to push and thereby set into vibration the spring, the slowest internal motion of the cylinder. How much energy would be deposited in the spring? For such a mechanical system, there is a general proportionality between energy and temperature (except under rather extreme conditions where quantum mechanical effects are involved). This is temperature on the absolute scale, counted in Celsius degrees from the absolute zero of the dark, empty vacuum. The answer is that the temperature of the spring* would be raised about one degree. At room temperature, about 300 degrees on the absolute scale, the spring is *already* quivering with several hundred times as much energy as this pulse would put into it. How could one be sure that anything was being detected, unless a truly enormous, and presumably very rare, pulse came along?

First, it helps to have two or more detectors that are separated by large distances and to pay attention only to those presumed signals that appear *simultaneously* at all the sites. These *coincidences* are much more likely to be external in origin, rather than produced by the ever-present bombardment of purely local disturbances.

Second, it is possible to detect *very short* signals, despite the presence of much larger thermal energies. In the first experimental device, the bar, when struck, would ring from some 15,000 periods, or about 10 s, before the sound weakened appreciably. That is how much time must elapse before any significant change of thermal energy can occur. Within shorter time intervals, the thermal energy stays fixed. Accordingly, a relatively small amount of energy, in a very short gravitational pulse, *can* be detected in the presence of this effectively *constant* energy background. There are, however, other kinds of random energy inputs (noise) introduced by the mechanism used to amplify the weak signals. And so it goes.

It is believed that the pioneer apparatus for registering gravitational waves could not have detected an energy input unless it was

*It is *not* the bar as a *whole* that is raised in temperature by one degree. To do that one would have to deposit an equal amount of energy in each of myriads on myriads of different kinds of oscillations.

large enough to raise the spring temperature by at least 25 degrees. (Recall that one solar mass of gravitational energy, radiated from the center of the Galaxy, raises the spring temperature by *one* degree.) Accordingly, the announcement in 1970 that signals were being received in coincidence at two detectors separated by more than 1,000 km—signals that seemed to originate near the Galactic center and arrived one or more times a day—hit the scientific community like a sonic boom. If it did nothing else, the news created a new field of science by stimulating the construction of gravitational wave detectors in numerous laboratories around the world. Now there are detectors in the original spirit, and there are detectors applying new ideas intended to considerably improve the sensitivity of the instruments.

One such idea has the ring of familiarity. A laser beam is split into two beams that are directed toward widely separated mirrors; after several reflections back and forth, they are recombined to form an interference pattern. Pure Michelson-Morley! But attention is now focused on the effect of a passing gravitational wave. It will produce a relative displacement of the two mirrors (the counterpart of tapping the drum) and change the interference pattern. With increasing separation of the mirrors, radiation of longer and longer wavelengths can be effectively investigated; the optimum separation is half a wavelength. This line of development leads to astronomical distances and the use of spacecraft.

It has been suggested that quasars contain black holes of enormous mass, which, during their formation, would have emitted gravitational waves with wavelengths of millions of kilometers. Suppose that a spaceship receives signals from Earth and relays them back. Measuring the round-trip time keeps track of the distance between the spacecraft and Earth. As discussed in Chapter 4 in connection with the red shift, the returning radio waves have a Doppler displaced frequency produced by the motion of the spacecraft. Suddenly a pulse of gravitational waves of extremely long wavelength moves through the solar system. That should produce a short-lived displacement of the spacecraft relative to Earth, which would show up as a change in the Doppler shift. Careful observations were made of the signals of the Viking probe that inspected Jupiter and Saturn but to no avail. The quest will continue with future deep-space missions.

The most immediately promising approach is the use of low-temperature techniques. A substantial reduction below room temperature is clearly useful in diminishing the thermal energy in the

Open-end views of gravitational wave detectors at Stanford University

spring, the background against which the gravitational signal seeks recognition. But perhaps more important is the ability—afforded at low temperatures by superconductivity—to make amplifiers that introduce very little noise and to improve the mechanical isolation of the massive cylinder from the noisy outside world.

Just such a program has long been under way at Stanford University in collaboration with three other laboratories. Instead of being suspended in its vacuum chamber by mechanical means, the cylinder is coated with a layer of superconducting material and then raised by nothing more substantial than a magnetic field. A detection device developed at Stanford uses a very light superconducting diaphragm that is mounted on the end of the cylinder. The diaphragm vibrates at a frequency that can be tuned to match the deepest tone of the cylinder. Then any displacement of the cylinder ends is transferred to the very light diaphragm as a much larger displacement; that motion is sensed by superconducting coils, which send a current to a SQUID.

Such instruments have recorded what might be interpreted as gravitational wave signals. But that identification will, to some degree, remain controversial until it happens that a supernova flares in the sky and simultaneous events are registered in several gravitational wave detectors around the world. Then shall mankind clearly have created a new sense organ, one attuned to the death and the birth cries of matter itself. The poet foresaw it:

And other spirits there are standing apart
Upon the forehead of the age to come;
These will give the world another heart
And other pulses. Hear ye not the hum
Of mighty workings in the human mart?
Listen awhile ye nations, and be dumb.

JOHN KEATS, "Great Spirits Now on Earth Are Sojourning,"
November 1816.

THE BINARY PULSAR

Radio telescope, Arecibo

It may be that the reality of gravitational radiation has already been proved, not on Earth, but in the heavens.

The search for new pulsars, using the 305-meter-diameter radio telescope at Arecibo, Puerto Rico, turned up something very unusual in 1974. A pulsar was found with the very short period of six-hundredths of a second. That is rare but not unique; the Crab Nebula pulsar blinks almost twice as rapidly. What was unique about PSR 1913 + 16, to use its cryptic astronomical name, was a disconcerting *change* of its pulsation period from day to day and even in minutes. It was quickly understood that the pulsar is part of a binary star system; it moves rapidly about its unseen companion with an orbital period of about *eight hours.* The to-and-fro motion of the pulsar produces the observed change in the pulsation period. The orbit is highly eccentric, the distance of closest approach—the *periastron*—being about a quarter of the largest distance in the orbit. The binary pulsar is much more relativistic than the system of the Sun and any one of its planets. If we use Earth for comparison, typical binary pulsar speeds are ten times as great, and distances a hundredfold as small. That should show up in the rate of *periastron precession.* It is measured as somewhat more than 4 *degrees* of arc per *year;* recall the perihelion precessions of Earth (4 arc sec/century) and Mercury (43 arc sec/century). But whether this periastron precession agrees accurately with the prediction of general relativity cannot be tested unless we know the total mass of the binary pulsar system.

The data accumulated in the years of continued observation have permitted other relativistic effects to be measured: the special relativistic time effect and the gravitational red shift have been isolated. That enables the masses of the stars to be determined individually; both are about one and a half solar masses. It turns out that the general relativistic prediction for the periastron precession agrees closely with the observed value.

Here is strong evidence that the mute companion is *not* a normal star; it must be small compared with a star like the Sun, because the size of the binary orbit is about the same as the diameter of the Sun. If the companion were as big as the Sun, the close proximity of the compact pulsar to the surface of its companion would distort the companion's mass distribution, raise enormous tides, and thereby change the nature of the gravitational force between them. That would produce a precessional motion of its own, which is *not* observed. We also do not observe any eclipsing of the pulsar by its companion during part of each revolution, as would be expected if the companion has such large dimensions (unless the orbit is being viewed face-on in the sky, but then we should not see appreciable variations in the pulsar period). It is most plausible that the companion is also a compact star: a white dwarf or a neutron star (the odds are in favor of the latter).*

Why is it important that the companion be small? Because only then can one expect the major source of energy loss in the binary system to be the radiation of gravitational waves. With a normal companion, the enormous tides would produce a significant conversion of orbital energy into *heat*.

After observations had been made for four years, it became clear that the orbital period of the binary system is shortening—by almost one-ten millionth of a second per orbit. That is not much, but in somewhat more than four thousand revolutions the effect accumulated enough that periastron was reached about a second earlier than it would have been otherwise.

---

*But there is a fly in the ointment. An optical observation at the Kitt Peak National Observatory, in 1979, tentatively identified the companion with a faint star; a compact star should be invisible.

The shortening of the orbital period means that the binary star system is *slowly* collapsing, losing energy. And the observed rate of energy loss *agrees* quite well with that predicted by general relativity. The evidence of the binary pulsar is impressive but still somewhat indirect; the discovery of other systems of this type would give valuable support. But nothing can replace the impact, and the promise of future knowledge, that the direct detection of gravitational waves on Earth would supply.

We are not finished with what the binary pulsar tells us. A pulsar blinks because it is spinning. Like a lighthouse beam, the beam of radiation that it spews out into space flashes by us (after a suitable lapse of time) because of that spinning motion. The pulsar is a gyroscope. And, because it is moving through the gravitational field supplied by its companion, its spin axis should change direction. (Recall that this is one of the effects that the gyroscope satellite experiment is designed to detect.) If the spin axis of the pulsar happened to be in the orbital plane, it should change direction by about one degree each year, and that would slowly alter the direction of the beam. (This effect diminishes the more the spin axis moves out of the orbital plane.) In fact, just such a change has been found. With sufficient study it should lead to a further precise test of general relativity.

It is remarkable how Nature aids mankind's groping toward an understanding of the universe. As we raise the level of our scientific skills and sharpen our artificial senses, fascinating new phenomena continue to appear, testing and challenging our growing comprehension of Nature's grand design.

FAREWELL AT THE FRONTIER

Einstein's attitude toward the creative aspect of science was altered by the enormous success of general relativity and by his subsequent endeavors. That change can be traced through the words of two lectures, twelve years apart, that he delivered in England.

Speaking about relativity in London—the year was 1921—he said,

> I am anxious to draw attention to the fact that this theory is not speculative in origin; it owes its invention entirely to the desire to make physical theory fit observed fact as well as possible. We have here no revolutionary

act. . . . The abandonment of a certain concept . . . must not be regarded as arbitrary, but only as conditioned by observed facts. . . . the justification for a physical concept lies exclusively in its clear and unambiguous relation to facts that can be experienced. . . . The general theory of relativity owes its existence . . . to the empirical fact of the numerical equality of the inertial and gravitational masses of bodies. . . .[5]

That is a physicist talking.

In the second lecture, delivered at Oxford in 1933, he began disarmingly with,

If you want to find out anything from the theoretical physicists about the methods they use, I advise you to stick closely to one principle: don't listen to their words, fix your attention on their deeds.

Then, after discussing the relation between experience and reason in science, he remarked:

If, then, it is true that this axiomatic basis of theoretical physics cannot be extracted from experience, but must be freely invented, can we ever hope to find the right way? . . . I am convinced that we can discover by means of purely mathematical constructions the concepts and the laws connecting them with each other, which furnish the key to the understanding of natural phenomena. . . . the creative principle resides in mathematics.[6]

At this time, as he had done for the preceding fifteen years and continued to do for the rest of his life, Einstein was struggling to construct a unified field theory, one that would unite electromagnetism with gravitation. The fundamental difficulty with this program is that no physical fact demands such a unification. (As a friendly critic remarked, "Let no man join together what God hath put asunder.") It was the absence of any physical clue that forced Einstein to purely mathematical speculation. In this effort, he was ahead of his time. We know that a unified field theory must also include fields that differ basically from those of electromagnetism and gravitation. Inspired by Einstein's example, valiant attempts are being made. This once solitary effort is now at the frontier. In the words of Michael Faraday,

Nothing is too wonderful to be true.

NOTES

1. **Helen Dukas and Banesh Hoffman, eds.:** *Albert Einstein: The Human Side* (Princeton University Press, 1979), p. 18.
2. *Albert Einstein,* Library of Living Philosophers, vol. 7. (1949), p. 272.
3. **Stephen Brush:** *The Kind of Motion We Call Heat,* vol. 1 (North-Holland, 1976) p. 286.
4. The letter is reproduced in *Gravitation,* C. Misner, K. Thorne, and J. Wheeler (W. H. Freeman and Company, 1973), p. 544.
5. **Albert Einstein:** *Essays in Science* (Philosophical Library, 1934), p. 48.
6. Ibid, p. 12.

SOURCES OF THE ILLUSTRATIONS

Drawings by Walken Graphics, Inc.

Facing page 1
Courtesy of AIP/Niels Bohr Library
Page 2 (margin)
The Granger Collection
Page 2 (bottom)
Courtesy of AIP/Niels Bohr Library
Page 3
Art Resource, New York
Page 6
Courtesy of the National Aeronautics and Space Administration
Page 11
Courtesy of AIP/Niels Bohr Library
Page 13
© Russ Kinne, 1979/Science Source/ Photo Researchers, Inc.
Page 14
Scala/Art Resource, New York, with authorization of the City of Bayeaux
Page 16 (left)
© Jerry Schad/Science Source/Photo Researchers, Inc.
Page 16 (middle)
© Grapes Michaud/Science Source/ Photo Researchers, Inc.
Page 16 (right)
Stephen Northup/Black Star © 1980
Page 18 (bottom)
© Bill W. Marsh, 1979/Science Source/ Photo Researchers, Inc.
Page 20
Eidgenössische Technische Hochschule, Zürich
Page 24
Scala/Art Resource, New York
Page 25
© Ted Streshinsky/The Stock Market
Page 36
Courtesy of American Friends of the Hebrew University, Inc.
Page 38
Courtesy of AIP/Niels Bohr Library
Page 42
Courtesy of the Griffith Observatory
Page 44
© Russ Kinne, 1968/Photo Researchers, Inc.
Page 45 (top left)
© 1980 Winston Scott/Courtesy of Hansen Planetarium
Page 45 (top right)
Gene Daniels/Black Star
Page 45 (margin)
Photograph by Patrick Wiggins/© 1980 Courtesy of Hansen Planetarium
Page 56 (margin)
Courtesy of AIP/Niels Bohr Library
Page 57 (top)
PHOTO CERN
Page 68
The Fogg Art Museum, Harvard University/Bequest of Edmound C. Converse
Page 72 (left)
© 1980 Flip Schulke/Black Star
Page 72 (right)
Associated Press/Wide World Photos, Inc.
Page 73
Richard Howard, Camera 5/Black Star © 1982
Page 88
Courtesy of Hale Observatories
Page 91
Courtesy of AIP/Niels Bohr Library
Page 92
Magnum Photos
Page 93
Courtesy of Southern California Edison Company
Page 106
Courtesy of AIP/Niels Bohr Library
Page 107 (margin)
© Parker/Science Source/Photo Researchers, Inc.
Page 107 (bottom)
Courtesy of C.D. Anderson/AIP/Niels Bohr Library
Page 108
Courtesy of Fermilab
Page 109
Erich Hartmann © 1969/Magnum Photos
Page 112 (left)
Courtesy of AIP/Niels Bohr Library
Page 112 (right)
Courtesy of Fermilab
Page 116
Scala/Art Resource, New York
Page 118
Courtesy of Brookhaven National Laboratory

Page 120
Courtesy of the National Aeronautics and Space Administration
Page 122
Scala/Art Resource, New York
Page 127
Courtesy of Dr. Francis Everitt, W.W. Hansen Laboratories of Physics, Stanford University
Pages 129, 130, and 131
Courtesy of the National Aeronautics and Space Administration
Page 147
Courtesy of Bausch & Lomb
Page 148
Courtesy of the National Bureau of Standards
Page 150 (bottom)
Courtesy of Robert F.C. Vessot/Smithsonian Institution Astrophysical Observatory
Page 152 (top)
Courtesy of the National Aeronautics and Space Administration
Page 154
Bildarchiv Preussicher Kulturbesitz, Berlin
Page 156
Courtesy of the Department of Defense
Page 158
Giraudon/Art Resource, New York
Page 160
© 1983 James Nachtwey/Black Star
Page 171 (top)
Scala/Art Resource, New York
Page 175 (bottom)
Courtesy of AIP/Niels Bohr Library
Page 180
TASS from SOVFOTO
Page 181 (top)
Courtesy of European Space Agency
Page 187
The Granger Collection
Page 189
© 1970 AURA, Inc., Kitt Peak National Observatory
Page 190
The Bettmann Archive, Inc.
Page 191
Dr. Fred Espenak/Science Photo Library/Photo Researchers, Inc.
Page 194
Courtesy of the National Radio Astronomy Observatory, operated by Associated Universities, Inc., under contract with the National Science Foundation
Page 198 (bottom)
Courtesy of the National Aeronautics and Space Administration
Page 202
Burndy Library/Courtesy of AIP/Niels Bohr Library
Pages 204 (top) and 205 (top)
Courtesy of the National Aeronautics and Space Administration
Page 205 (bottom)
© 1982 Dennis Brack/Black Star
Page 210
Burndy Library/Courtesy of AIP/Niels Bohr Library
Page 213 (top left)
Courtesy of AIP/Niels Bohr Library
Page 213 (top right)
Courtesy of Griffith Observatory/Robert Webb
Page 214 (margin)
© Mary Evans Picture Library/Photo Researchers, Inc.
Page 214 (bottom)
Courtesy of the National Aeronautics and Space Administration
Page 215
Uta Hofffman/German Information Center
Page 217
© Omikron/Science Source/Photo Researchers, Inc.
Page 220
Courtesy of Michael Freeman
Page 230
Courtesy of Hale Observatories
Page 231
Photograph courtesy of the National Museum of American History
Page 234 (left)
Courtesy of C. W. Francis Everitt, Stanford University
Page 234 (right)
Courtesy of Peter Michelson, Department of Physics, Stanford University
Page 235
Woodfin Camp & Associates

INDEX

Other Books in the Scientific American Library Series

POWERS OF TEN
by Philip and Phylis Morrison and the Office of Charles and Ray Eames

HUMAN DIVERSITY
by Richard Lewontin

THE DISCOVERY OF SUBATOMIC PARTICLES
by Steven Weinberg

THE SCIENCE OF MUSICAL SOUND
by John R. Pierce

FOSSILS AND THE HISTORY OF LIFE
by George Gaylord Simpson

THE SOLAR SYSTEM
by Roman Smoluchowski

ON SIZE AND LIFE
by Thomas A. McMahon and John Tyler Bonner

PERCEPTION
by Irvin Rock

CONSTRUCTING THE UNIVERSE
by David Layzer

THE SECOND LAW
by P.W. Atkins

THE LIVING CELL, VOLUMES I AND II
by Christian de Duve

MATHEMATICS AND OPTIMAL FORM
by Stefan Hildebrandt and Anthony Tromba

FIRE
by John W. Lyons

SUN AND EARTH
by Herbert Friedman